Blue Skies over Wuhan

Contemporary Chinese Studies

This series provides new scholarship and perspectives on modern and contemporary China, including China's contested borderlands and minority peoples; ongoing social, cultural, and political changes; and the varied histories that animate China today.

Christopher G. Rea and Nicolai Volland, eds., *The Business of Culture: Cultural Entrepreneurs in China and Southeast Asia, 1900–65*

Eric Hyer, *The Pragmatic Dragon: China's Grand Strategy and Boundary Settlements*

Kelvin E.Y. Low, *Remembering the Samsui Women: Migration and Social Memory in Singapore and China*

Jennifer Y.J. Hsu, *State of Exchange: Migrant NGOs and the Chinese Government*

Ning Wang, *Banished to the Great Northern Wilderness: Political Exile and Re-education in Mao's China*

Norman Smith, ed., *Empire and Environment in the Making of Manchuria*

Joseph Lawson, *A Frontier Made Lawless: Violence in Upland Southwest China, 1800–1956*

Victor Zatsepine, *Beyond the Amur: Frontier Encounters Between China and Russia, 1850–1930*

Patrick Fuliang Shan, *Yuan Shikai: A Reappraisal*

Selina Gao, *Saving the Nation through Culture: The Folklore Movement in Republican China*

Andres Rodriguez, *Frontier Fieldwork: Building a Nation in China's Borderlands, 1919–45*

Yuxing Huang, *China's Asymmetric Statecraft: Alignments, Competitors, and Regional Diplomacy*

Elizabeth A. Littell-Lamb, *The YWCA in China: The Making of a Chinese Christian Women's Institution, 1899–1957*

Yihong Pan, *Not Just a Man's War: Chinese Women's Memories of the War of Resistance against Japan, 1931–45*

Emily M. Hill, *Chiang Kai-shek's Critical Years, 1935–50*

For a complete list of the titles in the series, see the UBC Press website, www.ubcpress.ca.

Blue Skies over Wuhan

The Evolution of Environmental Protection Policy in Hubei, 1970s–80s

YUN LIU

UBCPress · Vancouver

© UBC Press 2025

All rights reserved. No part of this publication may be reproduced, stored in a retrieval system, or transmitted, in any form or by any means, without prior written permission of the publisher, or, in Canada, in the case of photocopying or other reprographic copying, a licence from Access Copyright, www.accesscopyright.ca.

Printed in Canada on FSC-certified ancient-forest-free paper (100% post-consumer recycled) that is processed chlorine- and acid-free.

UBC Press is a Benetech Global Certified Accessible™ publisher. The epub version of this book meets stringent accessibility standards, ensuring it is available to people with diverse needs.

LIBRARY AND ARCHIVES CANADA CATALOGUING IN PUBLICATION

Title: Blue skies over Wuhan : the evolution of environmental protection policy in Hubei, 1970s–80s / Yun Liu.
Names: Liu, Yun (Associate professor), author.
Series: Contemporary Chinese studies.
Description: Series statement: Contemporary Chinese studies | Includes bibliographical references and index.
Identifiers: Canadiana (print) 20250167271 | Canadiana (ebook) 20250167301 | ISBN 9780774870818 (hardcover) | ISBN 9780774870832 (PDF) | ISBN 9780774870849 (EPUB)
Subjects: LCSH: Environmental policy – China – Hubei Sheng – History – 20th century. | LCSH: Environmental policy – China – Wuhan Shi – History – 20th century. | LCSH: Environmental protection – China – Hubei Sheng – History – 20th century. | LCSH: Environmental protection – China – Wuhan Shi – History – 20th century. | LCSH: Environmental economics – China – Hubei Sheng – History – 20th century. | LCSH: Environmental economics – China – Wuhan Shi – History – 20th century.
Classification: LCC GE190.C6 L58 2025 | DDC 333.709512/12 – dc23

UBC Press gratefully acknowledges the financial support for our publishing program of the Government of Canada and the British Columbia Arts Council.

UBC Press is situated on the traditional, ancestral, and unceded territory of the xʷməθkʷəy̓əm (Musqueam) people. This land has always been a place of learning for the xʷməθkʷəy̓əm, who have passed on their culture, history, and traditions for millennia, from one generation to the next.

UBC Press
The University of British Columbia
www.ubcpress.ca

Contents

List of Illustrations / vii

Preface / viii

Acknowledgments / xi

List of Abbreviations / xiii

Introduction / 3

1 Mapping Hubei in China's Environmental History: Both Cursed and Blessed by Water / 12

2 Groping for Stones to Cross the River: Early Lessons in Three Effluent Pollution Cases / 36

3 Air Pollution and Soil Contamination: Voices of Protest against Industrial Pollution / 60

4 Struggles for Policy Implementation: Establishing the Environmental Agencies / 84

5 The Right to Pollute: Resorting to Cost-Benefit Calculations / 104

6 Updating Environmental Governance in Wuhan, 1986–90: Further Regulations for Blue Skies / 131

Epilogue / 161

Notes / 169

Bibliography / 193

Index / 202

Illustrations

FIGURES
1. Urban Wuhan / 20
2. Wuhan levee and flood monument / 25
3. Wuhan Iron and Steel Company / 32
4. Map of Dong Lake, Ya'er Lake, and the Fu River / 41
5. Dong Lake, Wuhan / 42
6. Map of Ya'er Lake, Echeng, Ezhou City / 47
7. Map of Fu River, bordering Yunmeng and Yingcheng / 51
8. Map of Daye, Huangshi, Hubei / 73
9. Map of Dawu, Xiaogang, Hubei / 76
10. Aerial views of Wuhan, 1985–89 / 133
11. Huangxiao River and Jianshe Avenue / 138
12. Noise pollution detectors on the street / 141
13. Yangtze River dolphin and finless porpoise / 159

TABLES
1. Flood incidents in Hubei / 21
2. Events correlated to Hubei's environmental policy evolution from 1952 / 132

Preface

This book examines the evolution of environmental protection policy in Hubei province from the early 1970s to the late 1980s. It is intended for scholars and students who are interested in environmental history, China studies, or East Asia studies, but some of its research themes are also typically covered in development studies and environmental economics. Many cross-disciplinary researchers are interested in China's environmental evolution, relevant industrial policy, and regulation analysis during the remarkable period of socio-economic transition that is the subject of this volume. I take Hubei as a case study to capture the evolution of environmental protection policy through the establishment of state agencies, including environmental offices and other government branches in the province.

The book illuminates the origin of a shift in environmental governance in Hubei from an agenda dominated by economic growth toward one involving more regulation, and in doing so it explores the early failures of subnational state agencies of environmental protection to regulate industrial waste emissions. It thus rests first on illustrating the relationship between local communities and the agencies as they responded to industrial pollution in urban and rural Hubei. It explains how this relationship unfolded during the 1970s and 1980s. My geographical focus is Wuhan, Hubei's capital city, which has often been compared to Chicago, particularly in terms of its cultural, economic, political, and historical significance.

An enormous number of contemporary reports document repeated conflicts in Hubei, all of which were connected with insufficient policy implementation to contain industrial pollution. One can also note that some consistent patterns and continuity in state policy had emerged for the regulation of industrial pollution and environmental governance in the province during the 1970s-80s, when it gradually started adapting a system of public crisis management to address the problem. Nonetheless, local records indicate that this system, though somewhat responsive, was quite disappointing, causing widespread public discontent.

My primary intent is to outline the development paths of environmental governance in Hubei, beyond simply presenting an argument about its gradual evolving progress, which could be characterized as rudimentary but with occasional signs of success and some promising results. The volume examines several high-profile cases of industrial pollution in both rural and urban regions of the province. Official or quasi-official records, all produced by government agencies in Hubei, show that pollution had a harmful impact on water, soil, and air, particularly in Wuhan. Some aspects of this have been largely ignored by earlier scholarship.

Applying the case study approach with archival evidence reveals how environmental policies were gradually adapted and implemented in Hubei. Additionally, a major motivation of my book is an effort to find evidence of environmental governance, contributed by civic groups or individuals. The case studies are well known locally but have not been adequately studied by external experts. My archival findings can contribute to a more inclusive analysis of China's environmental history. During the social and economic upheavals of the 1970s and 1980s, the public attributed ecological deterioration to administrative misconduct and policy negligence; yet some evidence of citizens' environmental awareness still existed.

This book contributes to a comprehensive interpretation especially on narratives that China's subnational state-policy framework consisted of both formal and informal information channels, with many offices designated to handle environmental protection. The topic of industrial pollution regulation and environmental governance is of major and obvious significance, given China's ongoing environmental problems. To date, no book has focused on this specific topic within a regional context, and so my volume offers some tantalizing glimpses of environmental protest and the government's responses to industrial pollution in Hubei from the early 1970s to the late 1980s.

Ultimately, my study reveals that government policy could not deliver results without strong support from both local governmental institutions and their corresponding civilian communities. Hubei's governance mechanism, consisting of citizens and bottom-level environmental protection agencies, responded to certain pollution cases that evolved with public discourses and sectoral conflicts, which would hit another climax in the 1990s. This evolution helps to explain how a symbiotic relationship – just like the complicated environmental governance system under examination – between civil society and state agencies shaped policy trajectories for the regulation of pollution until the present day, a direction that hopefully entails a more inclusive mechanism for environmental governance.

Acknowledgments

I owe thanks to a long list of professors whose guidance and support I highly value, along with the strongest assistance from my book editors and the valuable comments of three anonymous reviewers. My greatest gratitude goes first to, but is not limited to, the following faculty members in the Department of History at Queen's University in Kingston, Ontario: Timothy Smith, Rebecca Manley, Jeffery L. McNairn, Emily Hill, and Colin Duncan. I am extremely grateful for the valuable advice and encouraging words kindly offered by Andrew Jainchill and Jeffery Collins, along with many outstanding cross-disciplinary scholars and historians at Queen's, including Warren Mabee, Kim Richard Nossal, Allan English, Sandra den Otter, Amitava Chowdhury, Awet Weldemichael, Aditi Sen, Claire Cookson-Hills, Rob Engen, and many more. I must also offer my deepest gratitude to an even longer list of dear friends.

My revision process benefited from advice by my former colleagues across the History Department and the Department of Politics and International Studies at the University of Regina in Saskatchewan – Phillips Charrier, Jim Farney, Yuan Ren, Raymond Black, Yvonne Petry, Simon Granovsky-Larsen, Arjun Tremblay, Yuchao Zhu, Dongyan Blachford, and many others. My first field trip to conduct archival research in China was supported by a Timothy C.S. Franks Grant and a Dorothy Warne Chambers Memorial Fellowship Grant. I also received research funding from the academic publishing fund at the University of Regina, plus a book publishing grant at the Department of International Studies at Xi'an Jiaotong-Liverpool University, Suzhou, China, where I finally wrapped up the monograph revisions in collaboration with UBC Press. In particular, with the greatest support from two editors: Megan Brand and Randy Schmidt.

This list of acknowledgments keeps on growing, and I cannot express my thankfulness enough to those who helped a project that covered so many research themes. The book also contributes to the UBC Press series Contemporary Chinese Studies. It came out of a long journey from my early inquiries into modern China's economic policies with its industrial evolution, as reflected by the struggle of its iron and steel industry. Here, I benefited from the guidance of Randall Morck, Rolf Mirus, Vikas Mehrotra, Ryan Dunch, and many faculty members during my earlier years as a graduate student at the University of Alberta. I started the revision and reviewing work around 2019–20, right before the COVID-19 pandemic broke out in Wuhan, Hubei, the focus of this book. My family and parents have given me all the moral support needed to finish this work. Without their backing and that of all my dear friends, I could never imagine myself trekking so far as to see it in print. All shortcomings in this monograph are solely my own.

Abbreviations

CCP	Chinese Communist Party
DPMS	Disease prevention and monitoring station
ERJ	*Economic Research Journal (Jingji Yanjiu)*
EAR	Environmental assessment report
EPMS	Environmental protection monitoring station
HYP	First Historical Archives-Hangyeping Gongsi
HBNJ	*Hubei Nianjian (Hubei Provincial Yearbook)*
HPA	Hubei Provincial Archives
HEP	Hubei Provincial Bureau of Environmental Protection
HBSZ	*Hubei Shengzhi (Hubei Provincial Gazetteer)*
PRC	People's Republic of China
WMA	Wuhan Municipal Archives
WHNJ	*Wuhan Nianjian (Wuhan Municipal Chronicle)*
WHSZ	*Wuhan Shizhi (Wuhan Municipal Gazetteer)*

Blue Skies over Wuhan

Introduction

My initial intent for this book was to discuss the environmental policy challenges posed by rampant industrial pollution in the province of Hubei and its capital city, Wuhan, during the 1970s and 1980s. It eventually built into a multidisciplinary project after I realized that Hubei's long and rich history of environmental challenges deserved a more comprehensive and in-depth analysis. During the pre-modern period, the province continually struggled with flooding and draining wetlands for crops; in the modern period, its most pressing issue became industrial pollution, which was barely regulated, causing grave environmental problems from the 1970s. Chinese governments initiated a preliminary framework of environmental policy during the 1960s. In the early 1970s and the late 1980s, the central government at Beijing moved to tackle increasingly critical environmental issues incrementally; provincial and city governments had also started to set up administrative systems to handle industrial pollution in response to growing policy directives from Beijing.

My study topics thus invite a debate on how Hubei coped with its environmental calamities, particularly around Wuhan and the crowded urban areas on the Jianghan Plain, with their fertile rice paddies and water management issues. The book seeks first to illustrate how significantly Hubei's urban and rural societies and state agencies responded to pollution concerns and how relevantly policy adaptations evolved with public or quasi-public negotiation among state agencies, civilians, and academic communities. Hubei has long been a hub of commerce and knowledge, and the city of Wuhan, a famous site of modern China's industrializing experiments, is well known for its transportation and higher education in central China. One should factor in these particulars to interpret a series of industrial pollution investigations that occurred in urban and rural Hubei during the 1970s and 1980s. Because Wuhan was so important, Hubei received more political attention and administrative resources from Beijing than other demographically comparable provinces. Consequently, the evolution of its economic and environmental policies is more likely to feature both subnational and national characteristics.

A quick review of relevant research themes may help demonstrate my perspective. First, some preliminary signs of growing environmental awareness and grassroots protest preceded the generation of environmental policy during the 1950s and 1960s, when industrial pollution severely damaged water resources. However, policy measures were not put in place until the 1970s, when Hubei's newly established environmental agencies and other government branches started responding to some unrest, especially tied to rural fisheries and the consumption of contaminated fish by urban residents. Second, though environmental awareness may have been limited in Hubei, it certainly preceded awareness at the national level, as a list of provincial and municipal agencies sprang up first, with more generally issued directives from Beijing following later. Third, the "right to pollute" emerged as a compromise between the demands to contain industrial pollution and the enablement of economic growth, after efforts to regulate industrial emissions and to enforce environmental protection faded and the state-led market-oriented reform era from the 1980s started to disincentivize those previously state-funded collective projects.

This book contextualizes Hubei's contemporary environmental history as an evolving multidimensional process. Although it extends only to the early 1990s, I hope that it will provide a baseline for further studies of environmental-economic policies in China. Moreover, I stress a neutral interpretation of the evidence rather than a political diagnosis and will primarily apply a historical approach that focuses on both ruptures and continuities regarding the challenges of industrial pollution. Two facts can limit inquiries on the historical data: first, many of the relevant documents are still difficult to obtain, and their translated versions can be inaccurate or subject to political or ideological interference; second, China's vast hinterland, though seemingly much more intricate in terms of self-reflections or regional subcultural branding, has attracted less scholarly attention than its coastal regions.

Because this volume is grounded in a wealth of archival material, particularly at the Hubei Provincial Archives and the Wuhan Municipal Archives, a brief note about sources is useful here. The material includes the provincial, city, and county gazetteers, or *fangzhi*, semi-official compilations published periodically by many county-level and municipal authorities across China. Gazetteers contain official narratives on significant regional events, cultural, demographic, social, and economic changes, and short biographies of many well-known locals. The archives also hold investigative reports on critical issues by administrative employees at various levels, ranging from entry-level clerical staff at the sub-district or sub-county level (township or main street) to senior officials.

Also in the archives are government records on long-standing environmental problems, which appeared with growing frequency from the mid-1970s onward. For instance, if a government memorandum formally reported on pollution in the Fu River, which crosses various administrative borders, several reports would also be penned in Wuhan and adjacent counties in Xiaogan. The reports by both provincial and county officers regarding this polluted river, dating from the mid-1970s, also seemingly intended to shift the pressure for pollution control to lower-level government branches and county environmental offices.

Newspapers were also normally required to send their articles to the archives, and their coverage partly proves that environmentalist attitudes emerged in Hubei before Beijing began to issue its directives on pollution. Responsible for collecting and preserving these records, the Hubei Provincial Archives responds to requests for more specific investigation and publishing or even more sensitive archival record-based investigation missions designated directly by the provincial government or deputized by state agencies.

In addition to official records and news reports, a large body of source material was created by academic researchers. Research institutes and universities in Hubei have published considerable amounts of field research analyzing the environmental conditions in the province. Their reports help illustrate how the institutional sponsorship of investigation analysis was organized, since central authorities and provincial agencies facilitated investigations with tracking reports. Archival evidence indicates an increasing trend of inter-agency cooperation on environmental problems during the 1970s and 1980s. For example, the Hubei Provincial Archives holds a research report on air quality submitted by a small group of visiting American environmental scientists in 1979; it was followed by an investigation report filed jointly by researchers who also documented the escalating air pollution in Wuhan and a few other cities, including Shanghai and Guangzhou. Likewise, most universities and research facilities in the PRC played a role in influencing environmental policies. These civil institutions, subordinate to central or local governments, were also required to maintain archives of their research work and administrative records.

Historical periodicals published in Hubei are another group of primary sources. From the early 1980s onward, gazetteers (fangzhi) and chronicles (*nianjian*) were compiled regularly. The chronicles are organized as the collections of the Hubei Provincial Archives documents as the primary source. The archives of Wuhan and Hubei both established their own editorial offices as well and kept participating in those local chronicle projects. This book has benefited from their systematic selection priorities on regional pollution cases among official records, enabling me to trace the most significant events reported by either provincial environmental protection officials or other administrators. The editing formats and choice of subject matter in the chronicles and gazetteers remained consistent through time. They provide a bonanza of primary source material.

Given that Chinese studies should always factor in regional diversities or complexities, readers can take this book as a gateway to the environmental history of Hubei and Wuhan. It casts a critical eye on some narrow or biased selections of primary sources and the polemical views of some researchers. Here is a list of my key questions: What did Hubei's environment look like in the 1950s-80s, and how did it get that way? How did early scholarship explain the province's environmental problems and decision-making? How did environmental bureaucracies emerge in China? How is Hubei a useful case study of environmental bureaucracy? Particularly, how does a series of industrial pollution incidents reveal the weaknesses in the regulations of central environmental agencies, how were environmental protection policies first formulated and implemented in Hubei, and how was the logic of "the right to pollute" regularized and partly accepted with complicated intra-community negotiations?

Chapter 1 begins with a review of the rather sparse scholarly literature that addresses the evolution of the PRC's environmental policy during the transitional 1970s-80s. It then moves on to set the stage for my central topics, in terms of the environmental history of Hubei, in reference to the evolution of China's environmental policy-making and implementation at the provincial or county levels, and the interplay between citizens, provincial institutions, and national policy.

Chapter 2 presents three case studies on water pollution in Hubei. Produced by the virtually unchecked emission of industrial effluent, it caused enormous environmental damage in affected communities, enabled in part by repeated regulatory failures to contain it. Most incidents of pollution were initially exposed by the communities themselves, who raised the alarm to local staff at the Hubei Provincial Bureau of Environmental Protection and other government branches. The chapter examines how state agencies and communities struggled to address the problem, and how the logic of regulation eventually evolved into the solution of pricing "pollution rights" through closed-door or public negotiations with polluters.

Chapter 3 explores air and soil pollution in Hubei, including both urban and rural cases, to demonstrate the establishment and evolution of environmental institutions in the province. Investigative reports written at the time show that after incidents of severe environmental damage, officials took the initiative to gain control over the sources of pollution. Some officials, including those of the Hubei Provincial Bureau of Environmental Protection, were involved in mediation with victims and negotiation with officials in other state agencies. In some parts of Hubei's governments, feigned or genuine ignorance regarding official policies was a typical response to increasing environmental awareness.

Chapter 4 recounts the establishment of environmental protection agencies in Hubei, which can be traced to the early 1970s and the mid-1980s, to discuss how new institutional measures contributed to the achievement of environmental policy goals. Initially, there were many such agencies, but they had conflicting duties, due primarily to the administrative weakness of China's fragmented bureaucracy. A few survived this period to handle some aspects of pollution events and to codify environmental policy practices. Archival documents reveal a hidden policy priority to balance sectoral economic growth and monitoring duties of pollution regulators. However, a substantial portion of official responsibility for environmental protection remained in governmental branches that dated from the transitional period of the 1970s to the late 1980s.

Chapter 5 examines how provincial and municipal environmental agencies struggled with administrative red tape, underfunding, and understaffing. Hubei's environmental offices also demonstrated initiative and policy adaptation in implementing mandates to protect the environment and economic principles. This chapter explores how the practice of pricing the right to pollute evolved in Hubei and how local realities compromised national environmental policies. Here too, the weakness of the bureaucracy played a role within a set of provincially intertwined conditions. The chapter also underlines the contribution of a small group of Chinese economists in the 1980s, who provided vague guidance without specific information on how quotas and the right to pollute actually worked.

Chapter 6 concentrates on 1986–90, the period of the seventh five-year plan, scrutinizing the numerous reports published in the *Wuhan Municipal Chronicle* at the time. Cross-referenced with some early reports, the records in the *Chronicle* reveal how local environmental authorities adapted measures to the imperative task of regulating pollution. Some launched policy experiments that were consistent with pragmatic agendas. Some deliberately resorted to a sense of imperative existential threat or some narratives of doom, while shifting to a legitimization agenda to mobilize public action. These lessons in local governance underscore the public responsibilities that are shared within civil societies.

I would also stress that cultural factors shaped the trajectories of environmental governance by engaging a variety of actors that extended from state agencies to intellectuals and to everyday people, who often complained about the pollution that plagued their lives. Regarding the degree to which their input shaped industrial regulation and environmental policies, this book presents only inconclusive and limited evidence. Nonetheless, reports written by provincial or subprovincial environmental protection officers frequently mention public discontent, as a tactic to secure more funding and policy attention from their superiors.

This book aims to examine more locally oriented perceptions as opposed to overly simplified narratives, while providing caveats against resorting to binary arguments and "backward," or outmoded, labelling regarding the evolutions of the PRC's environmental governance. One may relate the following anecdote to the progressive realities and complex interplay between local environmental awareness and outsiders' mystified impressions. In the 1970s, an expedition team of wildlife field studies visited the Shennongjia area of western Hubei. Named after the legendary Yan (Flame) emperor, Shennongjia is the only forest district in central China, and many folklore stories in Chinese culture are set there. These field studies attracted public attention, especially regarding local tales of an ape-man or "wild" human-being who was said to live there. The expeditions found no solid evidence for the existence of the Chinese version of Bigfoot, which fuelled even more curiosity and sensation. Nonetheless, this mist-shrouded mountain forest resembles the location of many anecdotal accounts of numerous mythical beings. Such tales are unverifiable with little solid evidence – not unlike some similarly assertive remarks noted in my exploration of Hubei's environmental trajectory. One could argue that on many levels, Hubei is a representative case in China's path toward industrialization and that its difficulties in enforcing pollution regulation would resemble those of other provinces.

Many scholars attribute ecological decline in the PRC (not just Hubei) to misconduct and negligence in handling industrial waste emissions. Unlike the extant Chinese scholarship, my arguments highlight some specific details concerning how local environmental authorities adapted their policy practices during the 1970s-80s. Despite maintaining a primarily defensive tone, their official reports, archived in Hubei and Wuhan, demonstrate the efforts of subnational governments to grapple with the problem of pollution. Some officials undertook initiatives to gain control over administrative sources in response to a series of serious environmental accidents. Some, including agents of the Hubei Provincial Bureau of Environmental Protection, were also involved in mediation with victims and negotiation with officials in other branches of government.

Wuhan was somewhat privileged as it had a stand-alone government budget, which meant that its municipal agencies enjoyed some level of independence in administering their own economic and regulatory issues. This is confirmed by the material in the municipal archives, which reveals that Wuhan differed from the province. For instance, more air-pollution cases were reported in Wuhan than elsewhere in Hubei. In contrast, the records for regional water and soil pollution are typically housed in the provincial archives. In many cities across Hubei, municipal authorities stressed regulating air pollution caused by toxic industrial smoke. Back in the 1960s, the provincial offices and their county representatives had also undoubtedly noted growing signs of water-body pollution and soil contamination. This contrast in zones with high-priority issues reflects the preferences by provincial and municipal officials in local regulator offices or the public record bookkeeping systems in Hubei and Wuhan.

Archival records from the 1970s-80s help explain the failures in China's early efforts at controlling industrial pollution. A clear pattern emerges. Most tragic incidents of pollution were exposed, at least in part, during semi-public or public discussions. Complaining about effluent, urban smog, or contaminated soil, citizens demanded that something be done. But, very often, little was done. However, the distressing lack of results in environmental protection would prove to be highly relevant to Hubei's local governance structures, which underwent ongoing transitions involving many stakeholders – vocal or not. In fact, some local agencies were specifically mandated to defuse public pressure.

I conclude that the implementation of environmental policy featured a cost-benefit reasoning and inefficient governance. My inquiries help to interpolate how China's public governance and crisis management evolved with the state policy framework. Some policy experiments indicate that Hubei's societies participated in its governance mechanisms. Many policy practices entailed the growing participation of burgeoning civil societies. Increased efforts to be more inclusive help explain the policy trajectory. Many environmental governance issues are featured in China's local historical contexts, becoming part and parcel of the extended debate on global environmental governance evolving with new policy concerns. This book hopefully contributes to scholarly debates and public knowledge in a critical but understudied period of China's environmental history in an increasingly globalized world where, nevertheless, local societies remain deeply interconnected.

I
Mapping Hubei in China's Environmental History: Both Cursed and Blessed by Water

Given that the People's Republic of China (PRC) is a unitary state under central control, examining the relative strength and autonomy of its provincial government institutions is useful. A few relevant questions apply here: Did Hubei's environmental offices have the authority to enforce compliance with regulations and laws? Did any recognizable policy commitment from Beijing shape Hubei's efforts to control pollution? Examining the role of Hubei's environmental protection agencies helps shed light on these questions. If they continue to shoulder the bulk of the work for implementing national environmental policies, understanding their establishment and early evolution is essential in analyzing issues concerning the enormous pollution problems of today.

China is so vast and the existing scholarship on the establishment of its environmental protection agencies is so limited that applying most of the above questions to the national scale would be largely futile. Therefore, this book adopts a case-study approach to focus on a single province's early institutional preparation for the implementation of environmental policy. I selected Hubei partly because it is a significant region with certain political and economic advantages. Compared to other provinces of China, Hubei ranks in the middle of the pack for most socio-economic indices. For example, of the thirty-four provincial-level administrative units, it ranks fourteenth in land area and has the ninth-largest population and the seventh-largest economy.¹ Although it is not an affluent outlier like Jiangsu, Shandong, or Guangdong, it nonetheless holds a privileged position as relatively resource rich and politically and culturally influential.

Writing about public opinion on environmental policy practices, Peter Ho singles out a "developmental dilemma" in connection with land issues and China's institutional reforms. His analysis shows that private property with legal protection, the principle of "getting-the-prices-right," and emerging markets result from a given society's historical development or institutional fabric.² My findings support his view that the successful creation of new institutions hinges on policy choice and implementation timing, with a constellation of multidimensional parameters; disregarding this fact would essentially repeat the somehow clichéd narratives about poor stewardship and corruption grievances that are found in much English-language research on the subject. Also, Judith Shapiro further stresses collaboration between civil society and the Beijing government to shift China toward more promising environmental goals.³ Hopefully, growing confidence in China's cultural identity will help to create a much more predictable political culture in dealing with environmental challenges. Localized initiatives to maintain China's national intellectual heritage are evident in the writing of many influential scholars. Many local scholars have integrated history as part of the theoretical framework of environmental economics.

Many policy issues are tightly connected to a set of historical themes, as Philip Huang explores.⁴ Balanced views such as his examine modern China's ecological trajectories better than those that single-mindedly prioritize a few isolated factors. Here is a critical question that is hidden in modern Chinese studies: How well has the "public" space been established in China? As Huang notes, many inquiries address its third realm between state and society, and many questions evoke debates over key concepts, such as "public" versus "private," which is vigorously contested if vaguely defined. Plural institutionalism in China significantly shaped local policy preferences, and certain disagreements exist between a wide spectrum of intellectuals and the "party-state" (a politicalized term considered to be derogatory), or say, the central government in Beijing.⁵ Although the PRC's state-led economic reforms are often thought to have been launched in 1978, most post-Mao changes in industrial policies are rooted in a much earlier period. Such conceptual or institutional transitions, including restructured governance and dynamic policy evolutions, predated local pollution events, before a dramatic economic growth spurt after the 1980s, when pollution problems became more ubiquitous across China.

To justify confining my scope to the relatively short period from the early 1970s to the late 1980s, my historical approach stresses interpreting the language of local officials and their interactions with stakeholders. This approach is consistent with an idea suggested by J. Donald Hughes: environmental history should develop a method in which scientific and humanistic approaches are combined and should mediate between the two while paying attention to the stages of both cultural and natural ecology.[6] My culture-oriented inquiry is also inspired by the work of Elizabeth Perry, who brilliantly examines cultural features of China's revolutionary heritage from the early twentieth century.[7] My inquiries thus further advocate extending multidisciplinary themes in human geography, environmental law, and environmental politics.

Environmental issues become a hidden subject in many research fields related to the history of modern China. Many scholars drew on multidisciplinary perspectives more than firstly verifying archival sources. Mark Elvin deserves credit for introducing environmental history as an independent subdiscipline to the PRC during the 1990s; he proposed an outline research agenda and identified some specific themes worthy of investigation, expanding from interactions between various technical-ecological systems of water control to a systematic inquiry of the built environment in China. Elvin offers a word of practical advice: "the most promising method is to concentrate on the effects of the varying nature of the social foci at which decisions are made that affect the environment and the resulting feedback loops on society."[8] He further suggests that attention be directed to the evolution of China's state policy and its environmental impacts, as shaped by both official and non-official actors.

In a series of surveys, Maohong Bao discusses the environmental histories of both developing and Western countries.[9] Many similar reviews by Chinese historians draw on the pioneering works by Elvin, Martin Melosi on urban environmental history, and Donald Worster on the environmental history of the western United States.[10] As noted, China's public historiography is deeply interwoven with its environmental historiography and has also emphasized practicality while focusing more on the past than on the present. The reliance of Chinese scholars on state sponsorship for funding may compromise their objectivity for policy analysis, though regional studies look like an effective tool in Chinese environmental history studies.[11] However, integrating people's micro-stories and macro events would be consistent with the outlook applied in the "New West" approach practised in the environmental history of the United States.[12] Nonetheless, the small volume of China's environmental history writing stands in sharp contrast to the wealth of material that has become available in archival holdings during recent years.

Local narratives help contextualize specific circumstances that stakeholders have both confronted and adapted to. Regarding Hubei's ecological challenges, Jiayan Zhang examines the link between hydrological changes in the Jianghan Plain from 1736 to 1949 and county- or village-level responses to alterations in agricultural conditions.[13] According to Zhang, for centuries, Hubei's primary economic activities revolved around feeding the population that clustered on the plain. The principle of survival-first should be noted from local narratives in this "marshy kingdom." Zhang's work highlights the continuity-ruptures thesis in connection with China's environmental history. As one of the first environmental and socioeconomic histories of Hubei, it notes that the volatile environment of the Jianghan Plain presented constant challenges to local peasants who also must have adapted to frequent water calamities with rather ingenious and sophisticated responses over time. This book hopes to extend his examination scale with more multidimensional aspects.

Numerous articles analyzing China's environmental law help to explain the evolution of environmental policy after the 1980s, though I do not focus on impacts or failures of legal enforcement for environmental protection at the provincial level. During the early years of institutionalizing environmental protection, the PRC dealt with its pollution problems more administratively than through judicial processes. In the prolonged transition era, a typical development state paradigm started incorporating environmental policies into its legislative framework. Extending such legal inquiries into the political domain, Rachel Stern argues that even in China, where the law is at best ambivalent, environmental litigation can promote the roles of Chinese legal professionals in their daily practices while probing the boundary of what is politically possible.[14]

In an overview of China's environmental law, Tseming Yang summarizes ten facts to create a baseline for environmental protection legislation in the post-Mao period.[15] From the late 1970s, China began to implement wide-ranging legislation on environmental protection. Its environmental legislation was initially adopted on a trial basis in 1979, elaborating a general framework and set of principles for pollution regulation. The 2014 revision featured some significant updates, with new clauses on information disclosure and public participation while legitimizing the growing role of civil society in a system described as *yifazhiguo*, or "rule by law." China's judicial structures were more rooted in bureaucracy than those of other countries, despite some emerging distinctive features in its environmental law. Public litigation was still a novel tool; legal enforcement was a critical challenge, as an administrative responsibility that remained profoundly integrated within the bureaucratic hierarchy. Nonetheless, as Yang summarizes, "Divergence between requirements articulated in statutes and regulations and actual practice on the part of polluters and government agencies is a reality of the law anywhere."[16]

I also notice evidence consistent with Yang's perspective stressing environmental governance. As influenced by political tradition, this governance system is mostly decentralized in daily practices despite the PRC's tightly centralized control. As practitioners of national laws, subnational agencies must adapt to the fact that they are not working from a position of strength. In China's bureaucratic hierarchy, they often lack the authority to regulate state-owned polluters. However, an extensive regulatory infrastructure has been established since 1979. Yang notes that some observers have swung between predictions of imminent transformation and skepticism on bona fide progress. Conversely, positive signs of China's environmental policy evolution can support optimism for gradual but steady progress. Similarly, Paul Barresi argues that environmental policy implementation proceeds in particular ways in China, partly as its legal tradition is primarily based on Confucian meritocracy norms, which are not fully compatible with Western models.[17]

Issues of legislation enforcement for environmental protection, whenever at odds with higher-ranking administrative authority, are illustrated by Yuhong Zhao in his study on the resolution of disputes between polluters and victims.[18] Appropriate compensation for pollution damages has received more scholarly attention in recent years. Negotiating a compensation settlement still largely remains an administrative duty of local authorities. Michael Faure and Jing Liu note that the laws and government regulations were designed first to prevent and remedy environmental harms.[19] In alignment with such views, Adam Moser and Tseming Yang review China's 2009 Tort Law and its section on environmental torts within the cultural and historical context of civil litigation.[20] They note that plaintiffs whom, in many cases, are appealing for legal remedies for the pollution damage sustained, have enormous difficulties in collecting evidence, and they must often rely on their social connections to resolve civil disputes. If they wish to triumph in either civil or administrative courts, the victims of pollution must depend on experts from law firms, NGOs, local environmental offices, or relevant authorities.

Many scholars are interested in examining the political impacts of environmental challenges in modern China. Although they tend not to be historians, their first-hand work provides some essential guidance that was useful for this book. Vaclav Smil, a well-known Canadian geographer, is a renowned member of a small but growing group of social scientists who place environmental degradation in China within the context of official policy.[21] His path-breaking study, *The Bad Earth,* analyzes an outpouring of data that marked awakening environmental awareness and policymaking during the 1970s, together with a scathing indictment of the PRC's "mismanagement" of natural resources. Smil declares pessimistically that the best-case scenarios would result from "some gradual localized improvements and the prevention of further major degradations in key sectors and areas."[22] Selecting a town in Sichuan for his ethnographic case study, Bryan Tilt addresses public reaction and the political economy of weak pollution enforcement mostly at the village level after the 1990s.[23] His critical analysis and sharp observation of China's environmental governance at the bottom levels help us to understand the significance of China's rural-urban transition, which accelerated from the 1960s.

Elizabeth Economy contributes remarkable analyses of China's environmental protection policies; a political scientist, she examines the dire reality of policy constraints in achieving conservation goals after the 1970s. Using the case of the Huai, a river so polluted that it ran black, her survey also focuses on rampant water pollution from industrial sources.[24] Her work suggests some hints of local environmental awareness since 1972. Also concentrating on national-level policies, Abigail Jahiel examines the policy-transition significance of the PRC's Ninth National People's Congress, which announced a radical reform of central administration in 1998.[25] In an effort to scale down the bureaucracy, China's environmental administration was promoted to ministerial status. This reform looked like a bureaucratic anomaly at a time of strict administrative austerity after the 1990s. Many Western experts grounded their political analysis on specific policies and expressed pessimism about China's environmental policy implementation.[26] One can extend beyond such narrowly defined perspectives.

Most of the studies discussed above focus on national political changes, when limited experiences of participatory politics probably emerged in China.[27] Despite pessimism about the ongoing progress of environmental protection in the PRC, however, Richard Edmonds argues that there are grounds for optimism since environmental awareness and investment in keeping skies blue and water clean had increased with greater openness in Chinese societies, especially into the twenty-first century. Overall, China has presented a progressive path, with multiple venues that were seemingly viable for better environmental governance.[28] More scholars increasingly noticed that China's central and provincial governments and the public had responded to continual environmental challenges, though often differently. Nonetheless, there is hope for "state-led" environmentalism that arguably features both state control and public participation.[29]

A few more scholarly works deserve special attention and offer more motivating evidence for this book. Focusing on policy changes and citizen reactions, Andrew Mertha notes that in the PRC, water control and management has shifted "from an unquestioned economic imperative to a lightning rod of bureaucratic infighting, societal opposition, and open protest," as China has become increasingly market-driven, decentralized, and politically heterogeneous.[30] Implementing the relevant policies would involve a phalanx of state agencies, from the Hubei Provincial Committee of the CCP at the top and then down to various bureaus, ministries, departments, offices, and commissions, per their duties for economic planning, public finance, agriculture, forestry, and fisheries, plus a long list of industrial sectors. Examining relevant themes against the grain of political-ecological studies, Emily Yeh also regards natural resource governance as fundamental to rural politics in China.[31] I strongly agree with her view that we need to avoid "either exoticizing China or treating its rural [-urban] transformation as a tale many times told."³² My inquiry provides evidence of various bargaining efforts in Hubei to further extend some of the above arguments.

Introducing Hubei's environment, both cursed and blessed by water

Hubei is often depicted as a place of rich natural endowments and fascinating social customs. Located in the cultural core area of China, it has four distinct seasons, and some of its populated areas, including Wuhan, are subject to volatile weather. A local joke says that Wuhan can experience all four seasons in just one day: a spring-like and autumnal morning can be followed by a summer afternoon, and the day can end with a winter evening. Dramatic fluctuations in temperature interact with geographic features to produce a province that has an abundance of water.

This book focuses on a period of political turbulence, as much ideological confusion swept the country. As elsewhere in China, Hubei's landscape underwent enormous changes during this era. Judith Shapiro, who links this transformation to a "war against nature," presents a few case studies to demonstrate China's dire environmental situation from the 1950s to the early 1980s, though she does not discuss Hubei.³³ It is useful to conceptualize Hubei as both blessed and cursed in "[having] waters" and in "being [located] central[ly]," as a local saying notes.³⁴ Among its critical natural elements, such as water, soil, temperature, biota, and topography, the most important is Dongting Lake. One of the largest lakes in China, it lies to the southwest of Wuhan. In fact, it is so significant that the names of both Hubei and Hunan refer to it: Hubei means "north of the lake" and Hunan means "south of the lake." Over the centuries, the Dongting Lake area witnessed a rapid expansion of both agricultural production and population, a millennia-long trend that accelerated after the PRC was established in 1949, with subsequent decades of urbanization.

As the top decision maker of the PRC and the CCP, Chairman Mao was familiar with Hubei and Wuhan, and he often visited there after 1949. During his frequent stays or stopovers at his Dong Lake residence in Wuchang, he wrote many accounts of local and national events. Two well-known projects celebrate his status in transforming Wuhan's

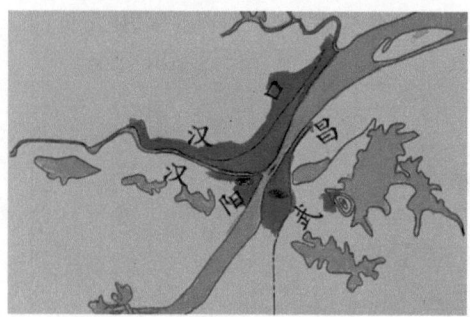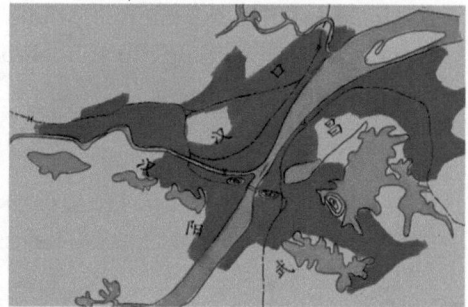

FIGURE 1 Urban Wuhan. The shaded areas in the maps show the urban zones of Wuhan in 1949 (34.7 square kilometres) and 1990 (189.3 square kilometres). | *Wuhan Municipal Chronicle* (1840–1985), Urban Construction Volume 1/2 *Wuhan Municipal Gazetteer* (Wuhan: Wuhan Municipal Chronicle Office, Wuhan Publishing House, 1991).

landscape. The first is the building of China's first bridge (the Wuhan Changjiang Bridge) across the Changjiang (Yangtze or Yangzi) River. Construction began in September 1955 and was completed in October 1957, finalizing the first national railway system to connect Beijing to Guangzhou. The second project was a new factory built on the outskirts of Wuhan by the state-owned Wuhan Iron and Steel Company, also in the mid-1950s. Both projects received loans and technical assistance from the Soviet Union. At his Dong Lake villa, Mao wrote one of his best-known poems, "Shuidiao Getou – Swimming." Celebrating anthropogenic changes in the landscape, this poem, first written in June 1956 but published a few years later, can be seen as the third monument to Hubei and Wuhan; it foresaw dramatic environmental changes and also the first evidence that Mao would endorse a mega-dam project on the Yangtze River.

As the capital of Hubei, Wuhan is as well-known as its lakes, rivers, and wetlands. The province of Hubei can be compared to a shod hoof: a semicircle of mountains – the horseshoe – rings its northern edge, the lower core area consists of a flat plain, and a notch opens at the south. The climate is primarily subtropical, with hot summers, mild winters, and infrequent frost, and the northern peripheral area has a mountain climate. The monsoon season commonly brings heavy rain. Seasonal floods often altered local watercourses, typically followed by the spread of pandemic diseases and food shortages in water-besieged cities and counties. In the early

TABLE 1 Flood incidents in Hubei

Century	Dynasty	Flood incidents	Average years per flood
2	Later Han (101–97 CE)	13	7.7
3	Later Han (219–98 CE)	11	9.1
4	Jin (301–99 CE)	12	8.3
5	Jin and North-South Dynasty (401–77 CE)	10	10.0
6	South Dynasty-Sui (501–600 CE)	No data available	
7	Tang (633–34 CE)	3	33.3
8	Tang (705–95 CE)	8	12.5
9	Tang (800–41 CE)	15	6.7
10	Five Dynasties-Song (925–1000 CE)	21	4.8
11	Song (1001–93 CE)	10	10.0
12	Song (1102–99 CE)	24	4.2
13	Song (1205–1300 CE)	22	4.5
14	Yuan-Ming (1301–95 CE)	44	2.3
15	Ming (1404–1500 CE)	46	2.2
16	Ming (1501–99 CE)	82	1.2
17	Ming-Qing (1601–1700 CE)	78	1.3
18	Qing (1701–1800 CE)	85	1.2
19	Qing (1801–99 CE)	97	1.0
20	Qing–Republic of China (1901–49 CE)	45	1.1

Source: "Flooding History in Hubei, Geography," *Hubei Shenzi* (Wuhan: Hubei Renmin, 1990), Table 12–1, 1185–87.

twentieth century, dike maintenance was poorly managed in Hubei. A flood in 1931 claimed hundreds of thousands of lives in Wuhan.[35] Flooding threat, particularly in the river basin that surrounds the city, is still fresh in the public memory of Hubei.[36] In fact, it dominated provincial records of natural disasters for nearly two thousand years, from the Later Han Dynasty in the second century to 1949 (see Table 1).[37]

Natural abundance and environmental calamities always accompanied each other – a paradox that has seemingly long existed in Hubei and likely other regions of China. According to the *Hubei Provincial Gazetteer*, about 70 percent of the croplands in the core region of the Jianghan Plain were devoted to rice paddies, whereas only about 30 percent were in the peripheral counties. Today, the population is still concentrated on the Jianghan Plain, despite the fact that it makes up less than one-fifth of the province's total arable area. This alluvial plain is criss-crossed by rivers and studded with lakes. The Han River flows into the Yangtze River at Wuhan; this flat fertile confluence is the most crowded area of Hubei. As the historical records demonstrate, frequent flooding was a reality in Hubei for centuries. With such a long tradition of flood memories, the farming communities should have deeply understood that water in Chinese culture can represent fortune, life, power, and the law of nature, but it can also bring destruction, disease, and death.

Prior to the Ming-Qing eras, the present-day jurisdictions of Hubei-Hunan were initially a single section of land unimpeded by waterflow before the Yangtze (Changjiang) River split them roughly into two provinces. After centuries of lake-river silting and dike-dam construction, the Jingjiang (the Jing River) section of the Yangtze became China's southern version of the Yellow River, with its population determined to fight the flooding caused by silty waterways. Ultimately, local authorities stepped in to take responsibility for flood prevention on the Jingjiang.[38] Some local scholars of historical geography argue that the plains on both sides of the river, including the alluvial plain of Jianghan and the floodplain of Dongting, were once part of an ancient, vast wetland called the Yunmeng Marshland, where rhinoceros and water deer roamed.[39] The product of this vanished swamp is one of Hubei's smallest counties, Yunmeng, in the jurisdiction of Xiaogan to the northwest of Wuhan. Many literary works retell the legends of Yunmeng from the pre-Qin period.[40] The southern portion of the ancient marsh became Dongting Lake, and a widely scattered chain of lakes marks other remnants of the vanished wetland. As a natural reservoir, Dongting Lake connects with the Yangtze River and with multiple outlets in Yueyang County and Yueyang City in Hunan. Since the Ming-Qing eras, the administrative maps and jurisdictions of Hubei-Hunan, with their winding borders, have remained largely stable.

The river-lake systems of the Lianghu (also known as Huguang, meaning Hubei and Hunan) region are fed by the Yangtze River and its tributaries. Most cities and counties along the Jingjiang River come under the jurisdiction of Hubei, except for a few near Dongting, which are administered by Hunan. On the whole, provincial borders have remained unchanged since the early Ming Dynasty (1368–1644), but the administrative maps of Hubei contain many zigzagging edges that may imply land infringement or evolving administrative convenience in managing irrigation or waterways.[41] These thin winding lines of provincial and city-county borders can also match topography, and they are largely aligned to the watersheds within or beyond the provincial boundary. Most wetlands that link directly to Dongting Lake on the south side of the Yangtze River are part of Hunan, whereas Hubei, which is famous as the "province of thousands of lakes," actually has a small portion of the largest lake in central China. Nonetheless, it possesses the most numerous lakes registered in the PRC.

Hubei has countless place names that begin with the Chinese word for "lake." As the lakes themselves no longer exist, these names are a testament to geographic change. Many place names end with *kou*, meaning "the mouth (of a river)," but these former river mouths are now completely silted up.[42] From the early Qing era, both Hubei and Hunan were well known for rice, cotton, and freshwater fish, and they shared the title of the "home of rice and fish," as well as their bordering plains.[43] Water scarcity long plagued the hilly regions on Hubei's western, northern, and eastern peripheries; floods posed substantial problems elsewhere. On the Jianghan Plain, the annual rainy season created persistent waterlogging issues. Irrigation solutions included constructing ditches and altering watercourses for timely drainage.[44] Protecting cropland from floods preoccupied the governments of Hubei and Hunan; the area of cultivation kept extending into the Yunmeng marshland. Over time, local communities adopted a series of land-reclaiming activities, either officially or tacitly approved by local authorities, via draining the wetlands and shallow lakes to create farmland to feed the growing population.

In 1952, Beijing launched a monumental project to control flooding on the Jianghan Plain and the Dongting Lake areas. Known as the Jingjiang Flood Diversion Project, or the Jingjiang fenhong gongcheng, its purpose was to divert floodwater from the Jingjiang River into a zone that would essentially act as a giant holding tank, thus relieving pressure on downstream areas. It came under the watch of Hubei premier Li Xiannian, who eventually chaired the National People's Congress (1983–88). In February 1952, Beijing summoned senior CCP and government officers from Hubei and Hunan to discuss the project. In March 1952, a directive titled "Decisions about the Jingjiang Flood Diversion Project" announced that the scheme would go ahead.

Getting under way in the spring and summer of 1952, the project was situated in Hubei rather than Hunan for two reasons. First, Hubei appeared to be more capable than Hunan of accommodating the diverted water; flood damage seemed more severe in Hunan than in Hubei because the floodplain in Hubei was at a higher level. Second, designing a diversion zone in an upstream area of Hubei was technologically viable, though its dike sections were better built than those in Hunan. The plain served as a natural reservoir for seasonal floods. After centuries of silting, its water bodies had been continually encircled to create more rice paddies and cotton fields. The flood control project was intended as an emergency measure, to be used in the worst-case scenario. Once the flood on the Jingjiang reached a certain level, sections of the river dike would be dynamited, allowing the water to escape into the diversion zone, much of it agricultural, which had been evacuated. This took the pressure off the downstream areas, but the inundated cropland could potentially remain submerged for up to three months if necessary.

There were whispers that the government's massive flood diversion design had sacrificed Hubei for Hunan.[45] Of course, the influx of river water would immensely inconvenience its diversion zone, to say the least, but Beijing saw this as an acceptable sacrifice. Undeniably, Wuhan would benefit hugely from the enormous potential economic losses in the upstream area. The premise of the project was that it would prevent Wuhan's urban dikes from being breached. Either way, flooding pressure would be greatly diminished for the cities on the plain. The plan would be activated once the river rose to the level of the benchmark flood in 1931, which has been designated as the critical benchmark signal to trigger the Jingjing Flood Diversion Project since its initial design stage.

In the long run, the strategy of repeatedly submerging cropland is neither economically nor environmentally sustainable. The project was activated in 1954 for the first time but was never reused, despite a close call during the century-record flood in the summer of 1998. Although the emergency protocol was prepared annually, it was only put into full stand-by operation from 1955 onward.[46] Flood response drills were institutionalized as an annual event that coincided with the rise of the river. Moreover, every hour throughout the flood seasons, water levels are measured from Zhijiang

FIGURE 2 Wuhan levee and flood monument. *Top:* The levee at Wuhan edges the Yangtze River. *Left:* Erected in 1969, a stone obelisk commemorates a catastrophic flood that struck Hubei and Wuhan in July and August, 1954, only two years after the completion of the Jingjiang Flood Diversion Project. | *Wuhan Municipal Gazetteer* (Wuhan: Wuhan Municipal Chronicle Office, Wuhan Publishing House, 1987, 1990)

to downstream Jingjiang River areas and broadcast locally and nationally. The fact that the river survived the 1998 Great Flood, almost unscathed, was celebrated as a triumph of Chinese socialism. Public awareness of environmental crisis would long exist with the hovering issues related to water management. In short, China's rapid urban growth, industrial expansion, and agricultural intensification have long posed a challenge to its management of water.[47]

Water brings both blessings and curses to local societies, while water mismanagement entails unbearable results well preserved in local memories. Some senses of environmental emergency would be deeply buried in the memories of Hubei. After the 1950s, local societies faced challenges with industrialization and grain production; soon the priorities of state policies were reflected in the seizure of land and water resources, by either the state agencies or local villages. Most county economies were designed to secure staple food supplies for cities. In late imperial China, the grain supply had become a priority for the growing population; the PRC had an insatiable need for more arable land. In Hubei today, mountain and hill areas account for 55.9 percent and 24.2 percent of its total area, respectively; the plain accounts for the remaining 19.9 percent.[48] Hubei is well known for its lakes, but its mountain and hill areas are less publicized and basically serve to secure urban-based industrialization. Wuhan consolidated itself as an industrial hub, turning the adjacent regions into subsistence-supply or satellite industrial towns.

In general, the organized conversion of Hubei's wetlands after 1949 took place on three levels: Beijing-led, by the army or government ministries; Wuhan-led, by the provincial government; and county-led, by county administrations. It proceeded much like a tango, advancing into drenched wetlands and retreating from them, in an effort to alleviate the rising population pressure by increasing the supply of arable land. Hubei's population nearly doubled from 25.81 million in 1949 to 51.44 million in 1988.[49] All large-scale projects to turn "unpopulated" swamps into state-owned farms had official approval.[50] Labourers were recruited from nearby areas to construct reservoirs to accomplish these tasks.[51] The loss of lakes and wetlands led to competition for sources of clean water. A remarkable example of local interests competing for the resource of clean water is that of Liangzi Lake; three subprovincial entities, the cities of Wuhan, Ezhou, and Xianning, eventually shared jurisdiction over the lake, which had belonged solely to Ezhou.

Calculating the number of lakes in Hubei is contingent on the definition of "lake," which is determined by the size and depth of the water body.[52] Both may alter during periods of flooding and drought, potentially adding or subtracting lakes from the list. Some recent public reports by the Hubei provincial offices admit that a substantial number of lakes, with an estimated total of 1,309 in the 1950s, fluctuating over time and finally dropping to 750 in 2014, ultimately vanished into farmland or fishery ponds.[53] Some scholars suggest that Hubei's total lake area shrank from 8,500 square kilometres in the 1950s to roughly 2,900 square kilometres in the 1970s, while the index for defining a lake also dropped from 1,000 mu to 100 mu, according to the lately adjusted lake standards.[54] Nonetheless, lakes in other provinces have also contracted or disappeared, and a nationwide comparison indicates that Hubei still has the greatest number of lakes in China.

Late-Nineteenth-Century Industrial Legacies in Wuhan

I want to review another significant aspect of the history of Wuhan and then draw some vignettes of its industrial heritage. Modern Wuhan is an amalgam of three old townships – Wuchang, an education-culture core; Hankou, a commercial district; and Hanyang, an industrial centre. In 1949, the three were merged to become the new provincial capital of Wuhan. Wuchang is famous for having ignited the Revolution of 1911 and for priming China's Republican era. It was also renowned for its schooling districts, gaining a national reputation in the field of education. Hankou, once the most commercially oriented of the three towns, retained little of its historical identity beyond its image of commercial affluence until its more recent revival. Hanyang was one of the few cradles of China's industrialization, where its first modern iron mill, the Hanyang Iron Plant of the famous Hanyeping Company, was erected in the late nineteenth century. After the 1950s, industrial Hanyang experienced a renaissance.[55]

As an irreplaceable aspect of China's early experiment with industrialization, the legacy of Hanyang can be traced to Zhidong Zhang (1837–1909), governor general of Huguang. During his tenure, China's first modern industrial cartel, the Hanyeping Company, was formed by an amalgamation of three enterprises, the Hanyang Iron Plant, founded in 1889, the Daye Ore Mining Company of 1890, and the Pingxiang Coal Mining Company of 1892.[56] Their merger created the most prominent iron-steel firm in East Asia at that time, besides its first modern arsenal, the Hanyang Arsenal.[57] After 1996, China became the biggest iron and steel producer in the world, a position it has firmly retained ever since.[58] After about a century of struggles, the PRC seems to have unequivocally realized its industrializing goal with some heavy environmental costs, as represented by the prioritization of its iron and steel industry from the very start.

The creation of Hanyeping was part of the Yangwu movement, which focused on industrialization. Spearheaded by the state and characterized by a top-down approach, the Yangwu reforms were led by a group of Confucian scholar-officers who implemented "self-strengthening" initiatives that advocated for industrial experiments from the late nineteenth century onward. Governance issues significantly affected the development of Hanyeping, whose early stages can be divided into three: the *guanban* (bureaucrat administration) period of 1889–96, the *guandu shangban* (bureaucrat governance-merchant management) period of 1896–1908, and the last period of *shangban* (merchant management) after 1908. The management of Hanyeping evolved from absolute government control to a compromise quasi-governmental control to nominal merchant control. A recurring problem existed throughout its life: impediments to corporate governance. In a December 1889 memo to Zhidong Zhang, another Yangwu reform leader named Xuanhuai Sheng (1844–1916) highlighted four key issues as follows: management, location, capital sufficiency, and inventory.[59] Sheng's foresight proved prophetically critical, as all of his concerns became a reality. Specifically, bad management turned out to be the biggest problem, as Sheng astutely recognized. This emerged in the decision regarding where to situate the iron plant: Zhang opted for Hanyang rather than Daye, which was about ninety miles downstream.[60] Technically, Hanyang was not an optimal choice, except for its proximity to Hankou and Wuchang.

Later, this choice of location was frequently criticized because of the high transportation costs entailed in production.[61] The site selection made the situation even worse: the mills were built on low wetlands to the west of the Jingjiang River levee. Constructing the iron mills required a large initial outlay of money and labour to strengthen the levee, drain the water, and raise the marshland. Repeated efforts were made to consolidate the soil, but poor foundations caused frequent accidents.[62] An outpouring of impeachment reports highlighted the faulty management of the enterprise.[63] An early case involved the impeachment of the county magistrate of Hanyang during the plant's construction. Another case revolved around a middle-level director named Lu, who was responsible for the acquisition of local mines. His associates frequently accused him of cheating, stealing, and cronyism. His case stands out from the rest, as it lasted the longest and produced more records. In the end, Lu was symbolically warned about his "reckless" behaviour, and Sheng insisted that he receive a penalty that he be docked some vacation time.[64]

Another early incident of poor management resulted when spending skyrocketed during construction. A progress report of 1891 stated that less than one-tenth of the planned construction was finished, but it had already cost half of the original investment of 1 million taels of silver, plus a loan of 200,000 taels from elsewhere.[65] Even worse, claims of financial irregularities were frequently made, but few records indicate that anyone was held responsible for these wrongdoings. The Hanyang Iron Plant finally announced its opening for business in September 1893, but no money was left to cover its operating costs.[66] Zhang had no choice but to consider private financing. He was reluctant to allow a state-owned business to be operated by private interest. The *guandu shangban* policy was a feasible compromise, advocated by Yangwu reformists, to utilize the private financing of merchants (*shangban*) under government oversight (*guandu*).

Zhang believed that Sheng was a perfect candidate to apply the guandu shangban policy. Beijing approved Zhang's transfer request, but the formal designation of Sheng as chief director of the Hanyang iron facility and its affiliated mines did not come until 1896.[67] In June 1893, one of Sheng's assistants, Tianwei Zhong, presented a memo to Sheng about the guandu shangban policy for the facility. Ambitiously proposing the creation of the biggest iron and steel syndicate in Asia, the memorandum covered a detailed list of forty topics that applied to it, ranging from corporate finance and public shares to profit distribution and regulations. It requested some renegotiable years of taxation waivers, along with protection from the harassment of local governments. Its thirteenth item stated, "*guandu shangban:* the government will be responsible for audits and guarantees, while merchants will be responsible for production, sales, corporate finance and employment."[68] Zhong's memo and records indicate that it was quite well received by decision makers.

Ineffective corporate governance appears to have been a critical factor in China's early industrialization experiments. There are still many puzzles in the history of Hanyeping. Not until May 1896 did Sheng present a new *zhaoshang batiao* (call for investment, in eight articles) after his official appointment to Hanyeping.[69] His personal control over the Hanyeping facility should have improved its management, as he could have his staff appointed to take direct responsibility for the daily operation of the mills without delays. Since 1891, even trivial expenditures, such as paying a Belgian engineer for reinvestigating mines, had to be approved by Governor General Zhang.[70] Such micromanaging reflects governance issues and bureaucratic inefficiency. Under Sheng's control, the mills started operating at least nominally like a commercial firm instead of a branch of the government.

But problems persisted. The chief executive of the Hanyang Iron Plant, Gongying Zheng, as Sheng's deputy on the board, resigned from his post in August 1897, after only eight months into his appointment.[71] In his letters, Zheng complained that managing the staff drained his faculties and energy. His health had deteriorated and he felt frustrated and powerless, so he hoped to "voluntarily" leave his position with Sheng's lenient understanding.[72] He implied that he could not control the staff who were assigned to him. His allegation was confirmed by Sheng's nephew, the standing chief executive, Sheng Chunyi. The latter reported that some branch factories were stealthily using the facility's resources and raw materials to produce and sell products on the side. In one case, a worker privately made two hundred iron seals for a tea shop and was caught red-handed.[73]

Zheng's successors faced a similarly difficult situation. Another impeachment report notes the increasing seriousness of management problems in Hanyeping, including those at the mines at Daye and Pingxiang. This secret report, forwarded directly to Sheng, recorded thirteen cases of corruption and included the names of suspects and the details of embezzlement. Again, no follow-up investigation was recorded, and the matter just faded quietly.[74] Archived records show that Sheng's deputies attempted to close various loopholes by increasing the frequency of staff rotation, ensuring that some key positions were shared by two individuals, and sealing cargo with special iron seals to prevent it from being switched with low-quality cargo or being crossed off the record while in transport. A monthly financial report system was also established in 1897 at Zheng's insistence before he finally resigned.[75] However, these measures did not halt the complaints of corruption, and the financial report system did not solve the serious problem of continually escalating operational costs in the factories.

Local records suggest that, had it been given more time, the Hanyeping parent company might have improved with technological development and conceptual adaptation. But time was not on China's side at the turn of the twentieth century, as tensions increased between it and foreign imperialist powers. The situation at the Hanyang Iron Plant deteriorated without fundamental change. In a secret report to Zhang in August 1900, at the height of the Boxer Rebellion, Sheng expressed great worry that China's industrial efforts would fail, as most mines might fall under foreign control.[76] Defeated in 1901, China was forced to pay a war indemnity of 450 million taels to the eight allied foreign powers. When the Russo-Japanese War broke out in 1904, the Chinese state was too enfeebled to step in or get involved in the conflict. Having won the war, Japan expanded its sphere of influence and concession area in northeast China. The deteriorating political situation eventually undermined Hanyeping's early efforts. Barriers to financing posed a long-term challenge for China's industrialization agenda in its modernizing trajectory.

Eventually, the Hanyeping Company gave up its long-term resistance to borrowing money from foreign banks, hoping to solve its financial predicament. Given its problems with corruption, foreign lenders kept seizing control of it to secure their investments. After the Sino-Japanese War, the Treaty of Shimonoseki granted foreigners the right to open and finance factories in China. The first foreign loan for the plant and the mines was signed with the Germans in April 1899, collateralized by the coal inventory and Sheng's personal credit.[77] A series of long-term loans ensued, mainly from the Japanese and collateralized by mines and machinery. Afterward, Japan secured an exclusive right to issue foreign loans to the Hanyeping parent company; in 1904, a loan with a thirty-year term was signed. The ore and coal sales were collateralized to the Japanese for debt repayment.[78] Later monthly reports show that the company predominantly shipped ore and coal to Japan. It had evolved from buying

FIGURE 3 Two views of the Wuhan Iron and Steel Company factory. | *Wuhan Municipal Gazetteer* (Wuhan: Wuhan Municipal Chronicle Office, Wuhan Publishing House, 1985, 1988).

raw materials for its iron production to selling its raw materials. This outcome deviated greatly from the original intent of the Yangwu movement, which was to self-strengthen China.

In hindsight, the failures of the company seem almost unavoidable. It continued its struggle into the Republican era before the Japanese invaded China during the Second Sino-Japanese War (1937–45). One can conclude that China's first industrial syndicate was anything but a success. Nonetheless, it did leave some valuable legacies, as the Yangwu reformers mobilized to adopt organizing mass machinery production, as well as advocating for a restoration of the dynastic order. When Chinese society began to embrace modern corporate concepts, as trialled in the Hanyeping case, it had no solid corporate governance structure that promoted the sharing of information or that rewarded innovation and efficiency, but it had plenty of governance failures and fraudulent behaviour that coincided with political chaos. By contrast, China's relatively steady political climate since 1979 has probably contributed to the success of its later reforms.

Despite its flaws, the pioneering experiment of Hanyeping marked a transition from local gentry-managed and family-based rural agricultural production to large-scale industrialization in urban centres that were already floating a monetarized economy. Some answers to those questions of how to survive would lie in the trade-offs between long-term and short-term goals of decision makers, and local cultural factors may create catalytic effects for accepting new ways of production. The example of Hanyeping helps connect the dots in the industrializing path of the PRC. Relevant issues of governance have kept evolving with policy adaptions to internal balance within the local contexts.

Environmental Awareness in Hubei
since the 1950s

Despite the scarcity of studies on the environmental history of Hubei, we cannot assume that environmental awareness was non-existent in the province. Nor should we attempt to "write back" a nascent environmental movement in China. Nonetheless, many civilian actors have contributed significantly to raising environmental awareness in Hubei, and they deserve more attention. Among them, various research institutes played a role in assisting decision makers.[79] Urban Wuhan requires a diversity of intellectual supports for its economic activities, such as those provided by the Changjiang (Yangtze) Water Resources Commission. Stationed in Wuhan, it is an intra-municipal water management agency that is part of the China Ministry of Water Resources. It was reorganized in 1950, and its early mission included the design for the Jingjiang Flood Diversion Project. More recently, it involved in the pilot hydro-electric dam project at Three Gorges and a low-dam project, Gezhouba, in Yichang.[80]

During the Cultural Revolution of 1966–76, many state-sponsored universities and research facilities were restructured or closed down, and some were permanently relocated to Wuhan, the regional administrative centre of southcentral China.[81] Removing them to the interior was intended as a protective measure against nuclear attack. Wuhan's educational institutes were relatively consolidated during this decade of cruel cleansing and power struggles.[82] The city suffered during the political bedlam, but it remained untouched by some of the chaos, partly because it lies so far from Beijing. Although the Cultural Revolution targeted intellectuals, local administrations managed to produce investigative reports on water pollution in Hubei from the mid-1960s. A series of field reports, overseen by the Hubei Provincial Committee of Science, issued warnings about severe industrial effluent pollution in local lakes.[83] This concern arose from China's massive machinery production reboot of the 1950s, when industrialization finally occurred on a national scale. Some scholars argue that China's recent industrial achievements did not take off suddenly from the early 1980s, suggesting that their origin can be traced to earlier decades.

Hubei's scholar communities also contributed to examining relevant policy adaptions from time to time. One would assume that most of these intellectuals acted as first-hand observers and responsible stakeholders in connection with environmental changes caused by either natural resource extraction or industrial waste emission. Many academic papers addressed environmental change in Hubei. For instance, the Changjiang Water Resources Commission started publishing the *Journal of the Yangtze River* in 1955, which printed reports on water management, river navigation, reservoir-dam construction, and geological-climatic investigations. Besides publishing regular papers on agricultural sciences, the journal prints articles offering policy analysis. For another instance, the *Journal of Geomatics* started in 1976 and was intended for a readership of both state policy makers and geomatics engineers.

Many members of Chinese elite scholars might uncharitably be labelled puppets of the state. Despite questions about their autonomy and integrity, local researchers played a crucial role in raising environmental awareness.[84] From 1976 to 1979, for example, several research expeditions to the Shennongjia mountains in western Hubei involved a large group of scholars in botany, zoology, geology, and other disciplines.[85] By 1981, dozens of scholars had proposed that China's first national nature reserve be created in Shennongjia. Thanks to its steep mountainous terrain, which formed a barrier to industrial activities, Shennongjia's natural environment and bio-diversity had been preserved. In interacting with government officials, academic communities played a leading supportive role in recommending, creating, and implementing environmental policy.

Conclusion

One could argue that Hubei is a miniature version of inland China, featuring an enormously uneven distribution of water across its space and seasons. Its rich records illustrate the long path in which agencies struggled to implement effective environmental policies. Like its sister province Hunan, Hubei is a major site of both pre-modern environmental problems (seasonal flooding and the conversion of wetlands to croplands) and modern environmental issues (industrial waste pollution).

2

Groping for Stones to Cross the River: Early Lessons in Three Effluent Pollution Cases

Three cases of effluent pollution occurred in or near Wuhan from the 1960s to the 1980s, presenting remarkable examples of the rampant pollution that beset the PRC during the period.[1] These three cases, at Dong Lake, Ya'er Lake, and the Fu River, demonstrate the pollution-regulation models in the early policy experiments of China and Hubei, as well as continuities in China's environmental governance.[2] These non-linear evolutions, which engaged both civil society and state agencies, accompanied waves of bureaucratic or juridical reorganization, and they deserve more examination than they have so far received.[3]

Despite differing administrative processes, many reports of pollution in Hubei, emerging in the early 1960s and extending beyond the late 1980s, reflect complex interactions and multiple changes in the policy priorities of local state authorities during this difficult period. Debates regarding policy priorities are connected to regulation challenges faced by state agencies, as can likely be inferred from a plethora of government and legal records.[4] In the case studies presented here, failures in regulation all resulted in a typical pattern: the site would be inspected, its operation would be suspended, damage losses would be settled, pollution devices would be installed, and production would restart.

My main argument underlines the experimental nature of environmental governance activities in Hubei. To cross this constantly shifting river while groping for stones, as the proverb goes, Hubei's state agencies responded to the appeals of affected communities. The widely known Chinese proverb, "groping for stones to cross the river," generally refers to crossing a river by feeling for a submerged but stable stone, stepping onto it, pausing to search for the next stone, and moving on. It also underlies some sorts of life experience or wisdom, the principle that stresses taking small steps first, not big risks or reckless moves, and ensuring that your foundation is steady before moving ahead.

It is worth highlighting that, intertwined with local societies, the governments of China have long played a key role in environmental management issues. Confirming the critical role of the state, historians note that many environmental issues, in a variety of regional settings, were also largely associated with popular support for imperial rule, even to the present day. My argument also stresses the role of local government in coping with the same old problems of environmental governance. China's technocratic issues engage both political and technical terms, besides practical barriers impeding environmental regulation. Working with newly released archival evidence, this chapter aims to extend the continuity-ruptures thesis and to enrich relevant debates.

In 1986, the central government at Beijing included Wuhan and thirty-seven other cities in its second batch of famous historical and cultural cities, which were deemed prominent because of their rich urban legacies. The second batch was less significant than the first, which consisted of twenty-four cities, such as Beijing, Nanjing, Xi'an, Guangzhou, and Suzhou, plus Jiangling (Jingzhou today) in Hubei, approved in 1982. Nonetheless, the legacies of Wuhan, from Hankou, Wuchang, and Hanyang, deserve in-depth exploration. Hankou and Wuchang have largely reclaimed their original identities, whereas the industrial legacy of Hanyang has shifted largely to the Wuhan Iron and Steel Company, also known as the Wugang Company or Wugang Group, which is based mostly in Qingshan, east of the Yangtze River. An industrial district that lies to the north of Dong Lake, Qingshan sprouted as part of the first five-year plan of 1953–58, which concentrated on industrial development.

Regarding Wuhan's local evolution in the economic-environmental history of the PRC, multidimensional explanations should start from its long-term progress and institutional adaptations. As for how local government institutions dealt with internal conflicts or external impacts, we need evidence-based interpretations that are not grounded in Eurocentrism. Many empirical explanations to the questions concerning China's economic-environmental governance evolutions are aligned with institutional theory, which stresses the significance of culture and a common belief in economic evolution. I agree that it would bring more insight to follow the "revisionist" views as typically applied by the "California School" of historians while arguing that progressive evolution spontaneously occurred in China.[5] Nevertheless, this case-study approach, aligned with many explanations based on institutional factors to address development questions, helps to tease out insightful and convincing details. Case studies create a diversity of explanations and specific interpretations. Cultural legacies and public school-education constitutions would also promote a historical conceptualization of Wuhan's urban history, and we can focus on Wuchang to explore its perspectives on governance. In Wuhan, some well-known universities provided special public goods, such as preserving cultural memory and knowledge mobilization.

During the 1950s and 1960s, one of the most significant events for Wuhan was the construction of the Wuhan Changjiang Bridge, also commonly known as the Wuhan First Yangtze Bridge. This key project finalized the Jing-Guang (or Beijing–Guangzhou) Railway, which was one of China's first and most famous trunk railways connecting north and south. The first 1,215 kilometres, the Beijing–Hankou railway (Jinghan railway), were funded and managed by a group of major shareholders, including Sheng and Xuanhua, between 1897 and 1906. The sectional construction of its southern portion, the Guangdong–Hankou railway (Yuehan railway), started in 1900 and ended in 1936. The last piece of this national project was the bridge on the Changjiang (Yangtze) River. Years later, a second bridge over the Yangtze River was built in Nanjing; a double-decker rail and road bridge, it was a scaled-down copy of the bridge at Wuhan. Its design has frequently been criticized as short-sighted because it does not allow the passage of large ships. The resemblance between these two bridges is a reminder that after Beijing severed ties with the Soviet Union in the late 1950s and military tension developed along China's northern border, the bridge at Wuhan was less celebrated than its counterpart at Nanjing, which took its place as a symbol of nationalist pride.

Eventually, tensions between China and the Soviet Union became so extreme that the Soviets withdrew their technological support. Nonetheless, their assistance provided Wuhan with valuable management and technician training and bridge construction experience. Today, the possession of the first bridge across the Changjiang River is one of its most valuable legacies, something to be celebrated in Wuhan. By removing a bottleneck blockage on the Changjiang River, Wuhan consolidated its status as a mega-city and transportation hub, connecting north China and south China. In addition, the Wuhan Changjiang Bridge categorically laid a solid foundation for establishing the Wuhan Iron and Steel Company at Qingshan. The company inherited the legacies of the Hanyeping Company after merging its Hanyang Iron Plant with the mines in Daye, Hubei, and Pinxiang, Jiangxi. Today, there are eleven bridges across the Changjiang River in Wuhan alone, a number that has earned it the title of the "Bridge City," which it shares with another well-known upstream city, the Chongqing Metropolis.

In the following three case studies from Hubei, local records indicate that public pressures were primarily directed at the Hubei Provincial Bureau of Environmental Protection (HEP). Chapter 4 will specifically address its institutional evolution and background as an auxiliary bureau under other committees or departments, to a fully established department with a more independent budget and its own personnel. By the late 1970s, the newly established HEP was mired in a deluge of public appeals and protests regarding pollution.[6] Some HEP officers took the initiative of responding to these concerns, with the result that the bureau produced numerous investigation reports, administrative directives, and policy proposals (which are now accessible at the Hubei Provincial Archives).[7] Despite their underlying defensive tone, these official records help to depict how Hubei's environmental regulations evolved from containing waste emissions to regularizing the penalties for pollution, which were hammered out during closed-door negotiations between the HEP agents and the polluter. This historical context, which is insufficiently studied, can reveal official responses to pollution issues in China. The three case studies of water pollution reflect some strenuous efforts by local governmental agents in Hubei, including HEP officers, to rouse public governance action.

Effluent Pollution, 1960s–80s

Reports of effluent pollution began to emerge in Hubei during the 1960s. From urban areas to rural regions, its major cause was always the same – industrialization. In 1964, for example, the National Department of Health, the State Planning Commission, and the National Bureau of Metallurgy issued a joint memo that recorded outrage regarding direct emissions of industrial waste.[8] They were a threat to public health, and many of Hubei's factory workers angrily voiced their discontent while struggling to acquire more help from medical sources. Four years later, in a 1968 memo warning about the industrial waste generated by iron-steel production in Hubei, the Hubei Provincial Department of Health filed an emergency public health notice.[9] A 1972 report by the Hubei Provincial Bureau of Water Utility and Irrigation noted that severe effluent pollution frequently affected local fishery production.[10]

These early reports could project official attitudes toward prioritizing the containment of industrial effluent.[11] Most of them seemingly tried to avoid ideological intervention and the narratives of class struggle that preoccupied China at the time. The existence of substantial pollution would stem from either unregulated or under-regulated emission of industrial waste, as well as poor regulation of household waste. The reports show that industrial waste attracted more attention than household waste. The former usually created denser pollution, and the negatively affected areas were typically larger than local jurisdictions; eventually, it led to appeals from the public, which tended not to be the case with the household waste. The official response would also evolve with regulatory protocols with industrial policies. Without incurring any costs save for a few *ex-post* ones to the polluting state-owned factories, any *ex-ante* decision of ignoring effluent regulation with incident tagging would sound economically rational. Government agents' acquiescence to pollution was seemingly a result of their cost-benefit calculations. Although they failed to contain polluting activities, their reports serve as first-hand evidence.

Dong Lake

Located in Wuchang, Dong Lake (Donghu or East Lake) is at the eastern edge of Wuhan. In the early 1950s, a variety of communities and institutions co-existed by the lake and included fisheries and vegetable farms. Many

FIGURE 4 Dong Lake, Ya'er Lake, and the Fu River, pollution sites in Hubei. | *Map by Eric Leinberger*

provincial government offices were at the corner of its western shore, and a few public parks and guesthouses appeared on its eastern shore during the mid-1950s. As mentioned, Mao Zedong had a residence by Dong Lake. From 1953 to 1974, he visited this villa forty-four times, about twice a year on average. His longest one-time stay lasted 178 days.[12]

FIGURE 5 *Top:* Sunrise on Dong Lake, in eastern Wuhan. *Right:* A sailboat race on Dong Lake. | *Wuhan Municipal Gazetteer* (Wuhan: Wuhan Municipal Chronicle Office, Wuhan Publishing House, 1987, 1988)

Many heavy-industry factories stood on the northern shore of Dong Lake, including the Wuhan Iron and Steel Company in Qingshan, the Wuhan Boiler Company, and the Wuhan Heavy-Duty Machinery Factory in Wuchang.[13] Other civil organizations by the lake would include a long list of education-research institutes, such as the Wuhan Institute of Hydrobiology at the Chinese Academy of Science, as well as Wuhan University. Many of these were on the western shore, and they accommodated many thousands of students.[14] A significant percentage of residents were university students. Many others worked either for municipal government branches or government-funded public-service facilities such as hospitals or schools. In addition, units of the People's Liberation Army had restricted zones by the lake. Adjacent urban and urban-rural areas were spread across these civilian and military districts. By the 1980s, the urban-rural areas, or "urban villages," had retreated to the remote southeastern shore, though they still supply fruit, vegetables, meat, and fish to urban Wuhan today.

In the later 1960s, communities at Dong Lake became distressed by its polluted state. At that time, it was a source of public drinking water, a situation that persisted until the mid-1980s, when the city transformed its water facilities into pumping stations.[15] By 1973, industrial waste was receiving more attention than the discharge of household waste, which had also become much more conspicuous with the increasingly dense population.[16] Two sources of effluent at Dong Lake were easily identified. The first, located on the northeastern shore, was the Wuhan Iron and Steel Company, which had been operated directly by the central ministry since the time of the first five-year plan (1953–57).[17] The second, a military map facility, stood on the southern shore and belonged to the Wuhan Military Administrative Region, later the Military Command of Guangzhou. The paper trails from the early 1970s make the pollution easy to trace. With an increasing investigation scope, reports noted pollutant accumulation in the lake, including a variety of phenolic compounds, arsenic and cyanide pollutants, and toxic heavy-metal elements such as zinc, manganese, copper, and chromium.[18]

The military map factory and the iron-steel plant were first accused of "negligence" in discharging effluent into the lake. Other polluters, as was suspected, would also release effluent directly into the lake, with minimum water treatment. Sewage pipes and ditches from dozens of factories funnelled enormous quantities of effluent into the lake. It had once been a popular spot for swimming, but by 1980 few locals would dare to take a dip, as it had become visibly polluted and smelly.[19] The issue received public exposure in the local media, and many government memos addressed it. The largest provincial newspaper, the *Hubei Daily*, and the largest city newspaper in Hubei, *the Changjiang Daily*, both published articles on effluent in Dong Lake in 1978 and 1979.[20] The newspaper reports pointed to another problem, the water hyacinth, an invasive non-native plant whose impenetrable mats choke out native plant species and reduce biodiversity. Because it has the capacity to absorb toxins and helps to contain rapidly spreading eutrophication, it had been introduced to the lake as a counter-pollution measure.

Both municipal and provincial government offices also repeatedly complained about the deterioration of Dong Lake.[21] In principle, the city administration had jurisdiction over the lake, but its pollution more immediately affected the provincial executive branches. With their offices located on the north bank of the Changjiang River, city officials seemed to lack the initiative to regulate a pollution problem on its southern side, where a phalanx of civil organizations and military units were located. The Wuhan municipal administration, seemingly unable to defy its higher-level counterparts, had to comply meticulously with provincial and central directives. Shutting down the iron-steel mill because of pollution sounded pragmatically impossible and politically suicidal. By 1979, Hubei had become the second-largest iron producer and the third-largest steel producer in China.[22] As a state-owned company, the Wugang Group fell under the jurisdiction of the central government, which would need to issue its approval of any solutions to problems caused by these factories. Similarly, civil authorities, like the Wuhan municipal administration, probably could not instruct the military units on how to perform their duties. In 2016, seven military regions devolved into five theatre commands. The central government has Hubei regrouped into the Central Theatre Command, for which Wuhan retains a headquarters status as the logistics base.

The pollution problem persisted at Dong Lake, with the result that during the 1980s, residents were greatly concerned about the safety of continuing to rely on it as a source of drinking water.[23] In 1985, a study led by the Hubei Provincial Department of Science and Technology presented compelling evidence of various carcinogens in the lake. More urgent calls pushed officials to deal with the deteriorating water quality in both Dong Lake and the river systems within and surrounding Wuhan.[24] This problem was an embarrassing one for Wuhan, as its official narratives proudly claimed that, from the nineteenth century, it had been one of the first cities in China to boast public running water.[25] Most water pumps were maintained by districts, and residents relied on a grid managed collectively by community groups. These operation units had groundwater pumped, sanitized, and then mixed with the public water supply.[26]

Like many communities by Dong Lake, all the universities in the nearby area operated their own groundwater stations to supplement the public supply from the lake. A petition for an alternative drinking-water source emerged in the 1970s, but change did not occur until 1985.[27] In that year, an internal memorandum classified as "secret" revealed that the Hubei provincial government had decided to prioritize the task of upgrading water quality for the Dong Lake universities. As a public welfare "gift" on the recently created National Teachers' Day in September 1985, provincial officials demanded that drinking water issues be resolved within two years.[28] The memo cited the upgrade as the priority on a long list of welfare issues troubling local universities.

Finally, the construction of two facility-expansion plants was entered into the budget managed by Wuhan. They would take water from the Yangtze (Changjiang) River which was cleaner than that of Dong Lake. This project, designed to replace the drinking-water plants for the universities and districts around Dong Lake, received a budget of 50 million yuan. The funds would come from three sources: 25 million were in the budget review from the Wuhan municipal government, another 15 million would come from the provincial government, and the remaining 10 million would be collected by the universities themselves. They had to apply for a budget allocation from their superintendent ministries. Some, such as Huazhong Normal University and Wuhan University, came under the watch of the Ministry of Education, and their budgets were tighter than those of their counterparts. Better status might exist for those overseen by industrial ministries, such as the Wuhan University of Technology or the Wuhan University of Water Management and Hydro-Engineering. In the end, university students and faculty amassed the remaining moneys through fundraising campaigns.[29] The budget-approval report added that the money was a one-time collection, and a water usage quota was allocated to the updated population statistics, so no other fees or fines would apply to the universities using the water in the future.[30]

Ya'er Lake

Ya'er Lake (or Yaerhu) means "Duckling Lake." Lying southeast of Wuhan, it was originally connected to Liangzi Lake. In the 1960s, low dikes were built on Ya'er Lake to manage fisheries or to create land for rice and cotton cultivation. Located at the point where urban Wuhan gives way to the suburbs, Liangzi Lake has also long served as a natural reservoir of the Changjiang River during the monsoon season. Within the multiple jurisdictions of cities and counties, Ya'er Lake was a source of freshwater for three subprovincial administrative bodies: Wuhan, Huangshi, and Xianning. Compared to Dong Lake, it was more of a border lake. These two lakes, Ya'er Lake and Liangzi Lake, had connections to each other and the Changjiang River before centuries of seasonal flooding silted up their outlets. Until the early 1990s, Tangxun Lake, another satellite lake also once connected to Liangzi Lake, was the largest urban lake in China. In local official accounts, an "urban lake" is defined as a waterbody that is administered solely by one city authority. Ya'er Lake was close to the "city-village joint-section," where rural and urban spaces overlapped each other, entailing a higher chance of neglect. Wuhan, which was more metropolitan than Xianning and Huangshi, enjoyed greater privileges. Xianning prefecture was primarily agricultural, where vast areas of bamboo forest were interspersed with rugged hills and remote mountains bordering the provinces of Hunan and Jiangxi.

An industrial conglomerate, the Wugang Group had factories in various sites that succumbed to pollution. The pollutants from its site at Ya'er Lake, for example, had become a significant cause of distress in nearby communities by at least 1975.[31] Local authorities reported that its deterioration was linked to accumulated industrial chemicals. In addition, after a rich ore deposit was found in the metallurgical corridor running from Wuhan to Huangshi – another site of factories connected to the Wugang Group – Ezhou seceded from Huangshi in 1983 and regained its municipal status like that of its next-door neighbours Huangshi and Xianning. This administrative regrouping was consistent with the Wugang Group's operational pattern of intentionally detaching itself from some of the pollution

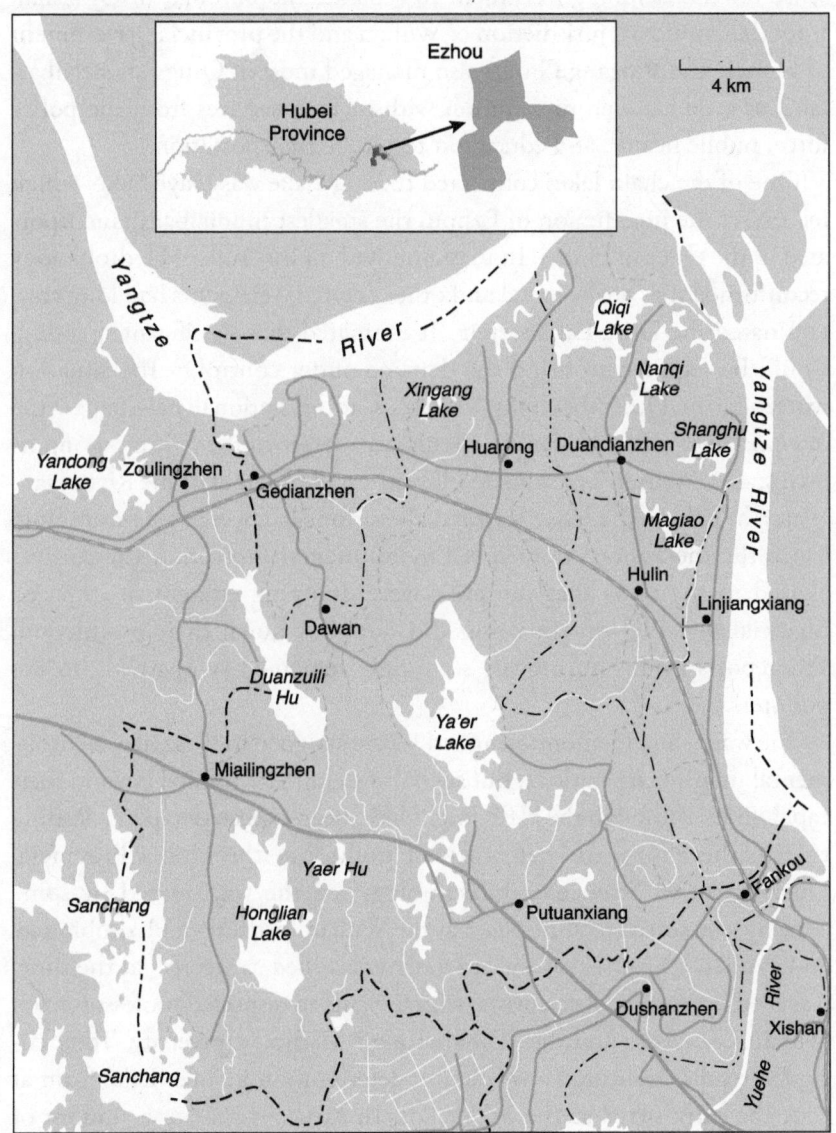

FIGURE 6 Ya'er Lake, Echeng, Ezhou City | Hubei Provincial Platform for Common GeoSpatial Information Services. *Adapted by Eric Leinberger*

problems it was responsible for generating. Similarly, although it made sense for the Beijing government to separate the borough of Qingshan from the municipal jurisdiction of Wuhan and the provincial government of Hubei, The Wugang Group also managed most of Qingshan's civil affairs, as a de facto administration with its civil services from the police force, public health, and education to public transportation.

One of the chain lakes connected to Ya'er Lake was Daye Lake, which fell under the jurisdiction of Ezhou, the smallest municipality incorporated in the 1980s in Hubei. In 1979, the Wuhan Institute of Hydrobiology reconfirmed that heavy metal and other chemical elements had long contaminated the Daye Lake system. It estimated that significant spending would be required to bring the damage under control.[32] The situation worsened: in December 1985, China's leading national newspaper, the *Renmin Ribao (People's Daily)*, circulated a letter to the editor among its own staff.[33] Written by a whistle-blower, a staff member of the River Dam Watch at Ezhou, the letter featured allegations of covert waste-dumping. Once the the Ezhou Municipal Environmental Protection Bureau was alerted to the news that the mills were dumping effluent, it imposed financial penalties on them but did not shut down their production. Yet, reports by Ezhou officers seemingly remained sympathetic to the polluters.[34]

The water-quality monitoring tasks were assigned to these new environmental monitoring stations, but such duties initially seemed beyond their capabilities. Although the HEP might have been expected to play a leading role in the efforts to keep track of pollution, the Hubei Provincial Department of Science and Technology led the way, joined by some education-research facilities such as the Wuhan Institute of Hydrobiology and Wuhan University. A set of reports, classified as "secret" at the time, was filed by a group of scientists, which included respected professors from Wuhan University and the Wuhan Tongji Medical University. Their co-signed reports indicated a worsening deterioration in Ya'er Lake from at least the early 1970s to the mid-1980s. In early 1970, a large amount of effluent pollution in the lake had caught the attention of central ministries, which quickly imposed pressure on local officers to contain pollution incidents.[35] The effluent seemingly came from chemical mills as well as the steel mills that had started operation much earlier. Central state agencies approved additional investment to update technology in the plants.[36] In the late 1970s, however, the factories were again accused of unlawfully emitting pollutants, leading to new rounds in which production was suspended and investigations were undertaken.[37] As all the factories were guilty of discharging wastewater, the local environment protection offices quickly executed temporary sanctions but soon alleviated them.[38]

Local concerns about public health escalated with complaints that effluent emissions had damaged fisheries in Ya'er Lake. Fish poisoned by the chemicals or heavy metals in industrial emissions ended up on urban dining tables. In a series of food poisoning cases in 1980, contaminated fish caused panic among urban consumers, creating enough attention for the desperate fisheries of Ya'er Lake.[39] After various attempts by municipal officials in Wuhan and Ezhou to assign the onus and blame to one another, local authorities in Ezhou promised to distribute immediate compensation to fishery workers and launched an inquiry to discover who had discharged the effluent. However, follow-up investigative reports indicated that these "scare tactics" had largely failed. Closed-door negotiations with mill managers determined what financial penalties and waste-process fees should be imposed on the industries, but once the fines and fees were collected, the factories simply continued to operate as before.[40]

After local authorities resorted to building low dikes to contain the effluent in Ya'er Lake, more complaints from lakeside villages, relayed by county authorities, flooded into the HEP offices. The dikes had created a new problem: In further separating Ya'er Lake from the main lake of Liangzi Lake, they helped to contain the effluent, but they also sabotaged the self-cleaning properties of the aquatic ecosystem. Such quarantine measures would be catastrophic for the water containment areas. Reports for Ya'er Lake show that the dikes had a negative impact on the rural side of the lake.[41] There were inconclusive results after constructing dikes to contain pollution on the northern shore of Ya'er Lake.[42] Yet the dikes seemingly created a buffer zone for the rest of the lake and prevented the pollution from spreading.

For the sake of resolving issues among local authorities, the subprovincial offices of environmental protection also requested feedback, first from the provincial government and then from Beijing.[43] However, the development goals of the iron and steel industry, to be delivered by these state-owned mills, seemingly outweighed local agendas for environmental protection. The Daye Iron and Steel Company (Daye Company/Group) – another syndicate of mining, iron, and steel – was also a successor of the Hanyeping Company, together with the Wuhan Iron and Steel Company.[44] Supervised by the provincial industrial agencies and the municipal government of Huangshi (then Ezhou), the Daye Company operated several mining sites across Hubei. Also owned by the Hubei provincial government, the Daye Group was a pillar of the local economy, as the largest employer in the city of Huangshi.

Adding to the complexity created by the iron-steel sector under the command of the National Bureau of Metallurgy, a large group of factories designated for chemical production operated under the National Chemical Industry Ministry. Many of them, located around Ya'er Lake, were state-owned: the Wuhan Gedian Chemical Factory, the Jian Han Chemical Factory, and the Wuhan Number 2 Chemical Factory, all around the town of Gedian.[45] The township was a convenient location as the hub of the chemical industry for the adjoining areas of Wuhan.[46] Besides the priorities of supporting state plans, local officials also faced a variety of branches of power that were directly responsible for industrial activities. Local offices, including the HEP and its subordinate offices, resorted to ad hoc measures to contain industrial emissions. They reported disappointing results in a new effluent investigation, as recurrently conducted at Ya'er Lake, a pattern similar to the almost concurrent pollution case in Dong Lake.

The Fu River

In 1979, an official investigation addressed pollution in the Fu River (or Fuhe), even as inquiries proceeded almost simultaneously with "renewed" investigations into Ya'er Lake.[47] Also similarly, the river served as a county border, with many sections under the joint bordering urban-rural jurisdictions. Regarding widespread contamination in rural portions of the Fu River, reports recorded more than twenty saline-product factories, operated by army units, provincial bureaus, and county authorities, discharging saline effluent into the river via a web of natural waterways and human-made channels. The military plants included those operated directly by the Hubei Provincial Military Region, the Regional Air Force Command in the Wuhan Base of the army, and other regional military logistical units scattered throughout the plain.

Three Effluent Pollution Cases 51

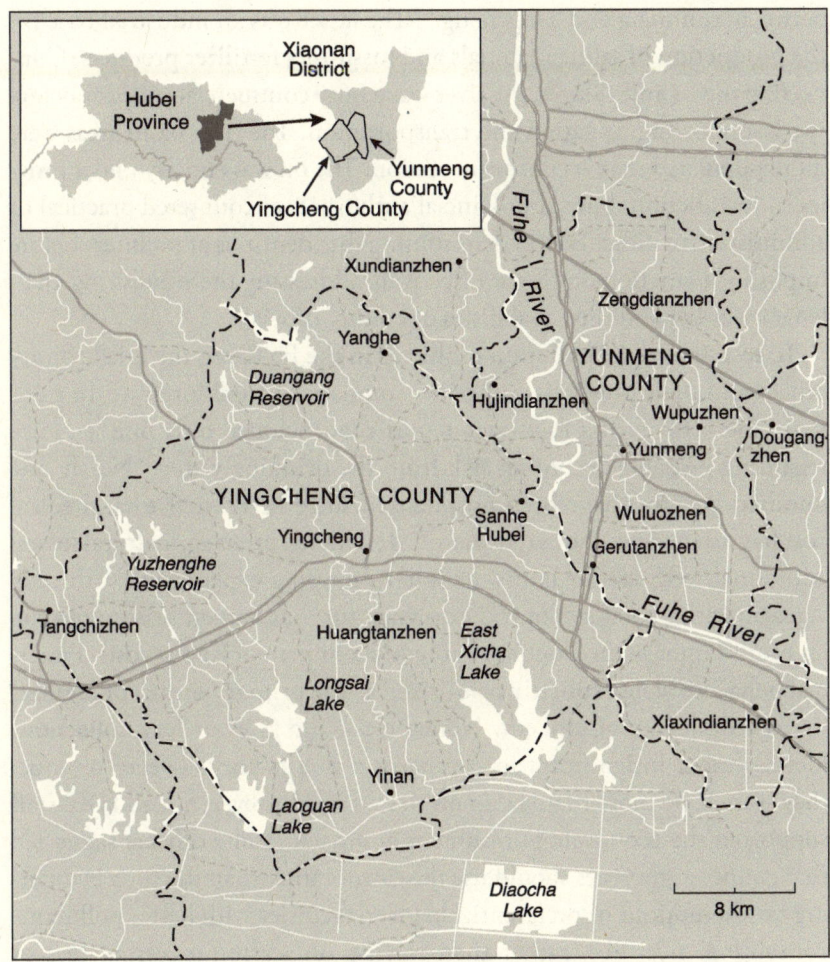

FIGURE 7 Map of the Fu River, which forms the border between Yunmeng and Yingcheng. | Hubei Provincial Platform for Common GeoSpatial Information Services. *Adapted by Eric Leinberger*

The Fu River winds through several municipal and county jurisdictions, from northcentral Hubei to the Changjiang River at Hankou. Irrigating the northern parts of the Jianghan Plain, as the Changjiang's largest tributary, and originating in the hilly areas of Xiangfan (renamed Xiangyang), the river stretches 385 kilometres through Xiaogan and Shuizhou before it reaches Wuhan. Also called the Yun River, it has long served as the border between Yunmeng and Yingcheng.[48] The latter was an industrial base for the production of saline materials and fossil-fuel fertilizer products. Connecting towns and villages, the river was a busy commercial corridor before 1949, supporting irrigation and transportation. The river basin areas were more populated than its upstream section. The Fu was entirely under provincial jurisdiction, but subprovincial authorities encountered practical or administrative difficulties in determining the identities of polluters before imposing a sanction on them. Effectively addressing these enduring challenges was beyond the capabilities of county officials.

To respond to the pollution problems in the Fu River, the HEP county offices and their provincial superiors emphasized rural interests. In mid-1979, the HEP filed a four-page report that included only one sentence regarding peasants becoming sick from the drinking water: "Nearly one hundred peasants have been hospitalized since 1978; their total medical cost has reached ten thousand yuan."[49] It detailed other agricultural losses: additional seed, labour hours, or water-pumping expenditure with crop acreage, with a total loss of 500,000 yuan for that harvest season, placing far more emphasis on economic rather than human or labour costs. Financial penalties were explicitly prioritized as the primary means of regulating the discharge of effluent. Nonetheless, the fines or levy collections looked rather insignificant in comparison to a factory's overall revenue; they were little more than a slap on the wrist. However, not all HEP staff supported the economic punishment approach. Some officers suggested that county authorities should use the carrot rather than the stick by offering tax exemptions or preferential loans to factories with better pollutant-containing measures, rather than relying on waste-processing fees or penalties. They also proposed that companies should be reimbursed for fines collected by county offices if they reinvested in pollution-abatement equipment and management.

Later, when responding to serious episodes of effluent pollution, national, provincial, and subprovincial officials turned to phrases such as "Explosive Public Hazard" and "Radical Response" to convey the severity of the contamination.[50] Their reports used strong words to stress the risk to public health, particularly when polluters were accused of repeated violations. They also pertained to continual efforts to negotiate between the parties in inter-sector economic conflicts. Compromises seemed necessary so that county authorities could maintain agricultural production at the same time and local bureaus could maintain the Wuhan Bureau's chemical production quotas.[51] Ironically, many saline chemical plants in Yingcheng were devoted to the supply of fertilizers and pesticides, together with other raw saline products primarily supplied to the agricultural sector.

Simply by using common sense, people in downstream villages could easily identify the effluent sources in upstream areas, many of which became the targets of blame. In their defence, however, one could argue that it would be unfair to censure a small group of factories, seemingly selected at random, while other violators avoided sanction. A few unlucky factories were scapegoated to deter other violators. The few publicly identified violators all belonged to a single category: they produced raw material and semi-finished products, affiliated with urban-industrial cities such as Wuhan.[52] Local officers of the HEP were put in the unenviable position of having to choose whether to defend agricultural interests or to protect the pollution-generating industries that were owned by the state.[53] Conscious of their newly established status, they cited agricultural interests as a primary reason for regulating industrial emissions. In this way, they avoided being caught up in a rivalry among industrial bureaus or confrontations with the patrons of industrial authorities.

In 1979, a closed-door symposium provided additional context regarding the failures in policy implementation. It was chaired by the vice-premier of China, Muhua Chen, who was visiting Wuhan during her inspection tour of food safety, epidemic prevention, and other environmental protection work. The Beijing government had already become alarmed about tainted food.[54] At the symposium, the provincial and municipal representatives reported difficulties in their daily work: understaffing, insufficient training in the use of pollutant-testing devices, and tedious protocols for access to some high-tech pieces of imported, expensive equipment. After the vice-premier encouraged greater use of advanced devices, a few officers sitting at the back of the meeting room spoke out, mocking the expensive equipment as nothing more than window-dressing since it was too expensive for effective deployment in rapidly growing pollution cases. Chen could offer no more than a frank admission that her hands were tied when it came to getting additional investment, recruiting more trained staff, or procuring new equipment.[55] Recurring effluent pollution came to the attention of Chen, who warned that new pollution cases would be subject to more public exposure.

In the case of the Fu River, importing pollutant-detection devices to quantify chemical residue accurately was assessed as unnecessary because there was so much effluent. Nonetheless, local officers of the HEP tried to circumvent red tape by industrial bureaus or local authorities.[56] The same cycle emerged through cooperation with military institutions: first, warnings would be issued; then, mandates were coordinated to suspend suspicious effluent discharge. This procedure would open a window of opportunity for the accused. After completing emission-control or technology-update projects, the plants received permission to resume production. Local officials of the HEP, along with other local officers, found it difficult to measure the damage, as they were given a predesignated time frame, with limited resources for assessing the impact of the effluent. Some waste inspections became mandatory, supervised jointly by other provincial industry bureaus, the military logistical units, and environmental officials in the (futile) hope of deterring polluters.

This pattern of joint regulation, coordinated among industrial regulators, local governments, and military authorities, was consistent with the handling of an earlier water pollution incident in Hubei. In 1973, industrial effluent from a military TNT arsenal in Yichen, Jingzhou, killed the crops of surrounding villages.[57] Managers at the arsenal blamed the lack of pollutant-abatement technology upgrades.[58] After a brief shutdown, the situation improved, partly thanks to the diluting effect of heavy upstream rain. Factory managers, their supervisors, and HEP agents declared satisfactory monitoring results jointly with new laboratory reports to support the conclusion of improvement. Nonetheless, local HEP officers had monitoring reports filed continuously about the Fu River pollution.[59] This déjà vu pattern dominated the loss-compensation negotiations for Ya'er Lake: first, officials would reimburse villagers while factory production was briefly suspended to allow technological updates or any pollution-abatement solution that was deemed cost-effective.

The HEP pollution reports on the Fu River also indicate that there were coordination efforts to establish pollution containment across administrative barriers. In almost every case reported by other state agents, the HEP officials typically delivered formal notes containing carefully phrased requests for visits by their inspection teams with county authorities, provincial bureaus, and military logistics units.[60] Predominantly administrative-economic measures existed for effluent regulation, including waste-discharge fees, administrative costs, waste-processing fees, or fines. The reports reveal that state agencies at the provincial level and below shifted from recording incidents of pollution to granting industries a de facto "permission to pollute." Disheartingly, the same scenarios kept returning to public or government attention, accompanied by more follow-up investigation reports by local environmental protection officers.

Reading against the Grain

After each round of pollution, detection, and amelioration ended, local HEP officials received new appeals regarding the under-monitored discharge of waste. Increasing site inspection followed closed-door negotiation for a fine or waste-processing fee. Although HEP officials claimed that efforts were being made to balance inter-sector growth, the reports reveal that manufacturing industries were benefiting at the expense of agriculture. Local HEP offices, prioritizing political-economic over eco-environmental goals, also paid closer attention to urban residents' protests than to those of villagers. Collective identities were often invoked in such protests, as observed in a few much more recent case studies.[61] In Hubei and elsewhere, rivers or lakeshores often served as administrative borders or jurisdictional boundaries, so wherever regulation duties overlapped, a space always existed for polluters to prevaricate, dodging costly deterrence measures.

Economic compensation varied for victims of pollution, a discrepancy that created resentment. For instance, gross calculation on salary and medical compensation, due to health injury caused by pollution, would be first divided into two house-registry systems: cities versus counties. The pay gap between urban and rural workers in China was huge, as were differences across regions. This disparity across urban-rural regions and economic sectors led to substantial variations in compensation, even in situations where the victims incurred the same damage from the same sources. It was unfair for those on the weaker bargaining side. The city workforce, throughout China, was mostly employed in state-owned industries or civil sectors such as hospitals, public transportation, and schools.

Compared to urban victims, villagers appeared to be the most innocent victims of illegal or under-regulated industrial waste discharge and were the least-compensated victims of pollution.[62] China's state authorities' economic-political foundation would shatter if the pollution consequences expanded.[63]

Hubei's environmental protection reports from the 1970s and 1980s illustrate subnational contexts evolving with policy priority in reducing pollution. Compromised enforcement measures and the local modus vivendi all contributed to regulation failures by state agencies, which are often criticized as passive or tacit players. Nonetheless, local HEP offices attempted to alleviate the impact of pollution, even though their policy measures all largely failed in the cases mentioned above. Regarding the protests of residents, direct evidence is limited, but the Hubei Provincial Archives contain numerous reports that refer indirectly to popular discontent. A passage from a 1970 emergency notice from the provincial government echoes these petitioning voices:

> Hubei has invested great effort in managing and utilizing industrial effluent, and has achieved positive results. However, some units have not sufficiently recognized this critical problem, paid little attention to effective measures, and consequently caused large amounts of industrial effluent to pour into lakes and rivers. In some regions, emission activities have affected fish reproduction, crop production, and even public health. Such phenomena require serious attention from the Revolutionary Committees at various levels.[64]

At times, higher levels of government exerted pressure on officials to contain industrial pollution for the sake of environmental protection. A memo of a 1973 speech given by Jun Yan, vice-director of the Hubei Revolutionary Committee (equivalent to the vice-premier), illustrates this situation in the provincial bureaucracy:

> Now there are some wrong viewpoints and foolish thoughts regarding the issues of environmental protection. Some said that our country is a socialist country; it is impossible that we [willingly] create pollution; consequently, we can simply dismiss pollution as a [duty-relevant] problem. What a foolish view!
> Some asked, "What chimneys do not discharge smoke; what factories do not discharge waste?" These staff members stand completely unconcerned …

Others said that there is not even enough time for production, so what time could be left to control three [industrial: water, airborne dust, and solid-state] wastes? These people do not understand that waste reduction measures will not hamper but promote production ...

Some, lacking confidence, grumbled about the hardship and emphasized getting more investment or equipment. This thought is also incorrect. It does require much time and effort to clean up the pollution caused by industrial waste, and this does require certain material conditions.[65]

Yan's critique targeted staffers who were dragging their feet when it came to tightening regulations on industrial waste. Some responded with negotiation in exchange for stricter regulation, such as funds for new equipment or staff. Reading between the lines in the reports, one also senses tremendous frustration. Environmental protection state-agents in Hubei seemed to know that their audiences, including both the central and provincial governments, should act more assertively to respond to the outcries of local communities. To some extent, they were responsive to public calls for quick intervention. Unfortunately, their routine measures, such as issuing warning directives, undertaking official inspections, and offering pollution-containment guidance, did not create any solid results or alter the fact that the manufacturing sector was favoured over its agricultural counterpart. Dire conditions in both urban and rural spaces continued to threaten the public interest, especially when state-owned industries received policy priority over farming.

The HEP's challenge was its limited authority to regulate both urban and rural pollution. Discrepancies in policy priorities elicited dialogue among state agencies, factories responsible for pollution, and local communities. When HEP officials started engaging various industrial interests, administrative measures, and state priorities, interim benefits outweighed long-term vision in policy adaptation. Handicapping critical matters included practical questions of how to measure economic loss, how to distribute reparation, and how to trade off benefits for urban workers and rural residents.[66] When evaluating those liabilities economically, investigative reports categorically recognized the need to maintain production routines for state-assigned quotas. Traces of continual struggles surfaced with these official inquiries, as realistically preserved in Hubei's local and provincial archives.

Practical policy stakeholders should learn how to sensibly adapt to such realities. Hubei's pragmatic approach to its industrial experiments is demonstrated in the following anecdote about its rich culture of stone/jade: Hubei has the largest turquoise reserves in China, supplying many far-flung areas such as Tibet. This semi-precious mineral, also known as green pine stone, is claimed to be used for China's first Imperial Seal from the Qin Dynasty (221–206 BC). This Heirloom Seal of the Realm, symbolizing the Mandate of Heaven, is recorded in myriad famous events in ancient China. All jade must be polished meticulously to retain its aesthetic value. Yet, as we wade across the river while groping for stones, we must choose them solely on the basis of their weight and stability if we wish to prosper. Turbulent waters do not distinguish between "valueless" stones and precious jade or treat people differently, as the power of nature disregards all human-made criteria.

The cases of pollution in Dong Lake, Ya'er Lake, and the Fu River all exhibit a similar pattern, and connect various causes that help explain the failure of Hubei's environmental protection policy. Their stories all link the past to the present and provide valuable policy lessons from Hubei's early attempts at regulating industrial effluent: more civil society engagement, but less external intervention or ambiguity of administrative or judicial accountability. Hubei's state agencies were aware of the negative impact of industrial emissions. Yet, a consensus still existed: the rampant pollution eventually led to inequitable economic costs. Occasionally, local HEP officials, caught in the cleft stick of having to choose between the agricultural and manufacturing sectors, raised their voices to capture more attention from their government peers or superiors regarding economic infringements caused by industrial waste. The challenges of mediating policy conflicts, involving so many executive branches and industrial sectors, lay beyond the scope and the remit of local environmental protection offices.

Conclusion

From the 1960s to the 1980s, communities in Hubei strove to control industrial effluent but without much success. Archival records reveal several reasons for the failure of early environmental protection policy, including administrative restructuring, power structure complexity, overlapping jurisdictions, and the prioritization of industry over agriculture. Shifting political boundaries provided convenient cover for neglect or fraudulent acts. Rising environmental awareness clashed with feigned or unwitting ignorance. Local resistance to external intervention, including pressured interference coming from either higher state authorities or any players identified not as members of local societies, should also be recognized in these overarching narratives.[67] Policy conundrums constantly dominated regulation practices at the provincial level and below. However, the effects of actual regulation were poor, as incidents recurred with rising public health concerns, entangled with cross-sector conflicts. Legitimation of administrative punishment, institutionalized later by the National Administrative Litigation Law, was not formalized until 1989. In that year, the HEP started resorting to court action for environmental policy implementation.[68] Such findings are consistent with the view that local courts helped nudge along incremental social change.[69] Even today, as Rachel Stern's analysis reveals, the way in which a court battle over pollution ultimately plays out still depends largely on local circumstances in China.[70]

3
Air Pollution and Soil Contamination: Voices of Protest against Industrial Pollution

Air and soil pollution around Wuhan was as serious as that of nearby lakes and rivers. Here too, local voices evolved in public or semi-public discourses on environmental governance issues during the 1970s and 1980s. As indicated in some recent policy studies, many practical difficulties of subnational governance for environmental protection were common in China.[1] A few 1960s reports regarding pollution in Hubei bear this out, providing hints of how local HEP offices initially responded to public concerns about governance failures in environmental protection. During the 1970s, reports of pollution increased in both cities and counties. Despite their limitations and defensive tone, many environmental protection officials in Hubei assiduously recorded scores of protesting voices into the mid-1980s. These barely explored records help us to interpret how official and quasi-official responses, especially at the provincial and subprovincial levels, dealt with protest regarding pollution during the 1970s and 1980s.[2] Capturing the "insignificant" protesters' voices in the reports of environmental protection officials also helps make connections within China's evolving environmental situation.

Air pollution was more evident in urban Wuhan than in the countryside, where soil pollution posed a more direct threat to villagers. Detecting air pollution was often a simple matter of looking up – at the obvious yellow-black smoke-dust that billowed from tall factory chimneys. Identifying contaminated soil in rural areas was considerably more difficult. In discussing air pollution, I focus on three districts in Wuhan: Hankou, Wuchang, and Qingshan. As mentioned earlier, Hankou was traditionally a commercial and municipal administrative zone, Wuchang was associated with higher education, and Qingshan was a national base for iron and steel manufacturing. Like Wuchang, Qingshan also contained newer factories. My discussion of industry-generated soil pollution concentrates on two counties: Daye, on the south side of the Yangtze River, and Dawu, to the north of Wuhan. Daye is proud of its industrial legacy from the late nineteenth century, and Dawu has long celebrated its revolutionary tradition with the Chinese Communist Party (CCP) since 1949.

To detect traces of long-term evolution in a single locale, changes in its name help reveal evolving social customs or historical status. Learning how places emerged also helps to contextualize environmental history. Accounts of place names in Hubei's gazetteers illustrate gradual and long-term environmental changes in the province. Many place names in the central plain regions have linguistic connections to water resources, and some of their rising and declining statuses were also connected with the evolution of their place-name stories. Shuffling names administratively on the map often confuses outsiders and scholars, which can be illustrated by the case of Wuchang. After using the name of Wuchang for 1,600 years, the city previously called "Echen" was stripped of its original title around 1912, and then the name of such a rich history was bestowed on Wuchang's annexed county, Jiangxia, as a memorial event to celebrate the Wuchang Uprising in the Revolution of 1911. Eight decades later, in March 1995, an even earlier place name, Jiangxia, was restored as a county-level borough of Wuhan. Restoring Jiangxia was a gesture of historical legacy restoration to the borough. Likewise, Wuchang lost its early names of Ezhou and Xiakou. This short history of the place-name changes of Wuchang indicates its pivotal position continually featured as the regional administrative and cultural centre, after the name "Wuchang" was shuffled among various nearby cities. The city of Wuhan has now monopolized the name of Wuchang as its unique urban entity and no longer shares it with other administrative bodies.[3]

Changes in place names may be indicative of an economic-political shift. A similar name grouping occurred with the city of Hanyang. Hanyang city and Hanyang county co-existed until 1992. The second character in Hanyang's name, yang, means either north of the water or south of hill ridges and refers to the city's original location on the north side of the Han River. At some point before the mid-1800s, the river shifted its course, with the result that Hanyang was now on its southern bank. Hankou, officially inheriting the name of Xiakou, emerged from numerous seasonal floods after that time.[4] After 1949, it was gradually eclipsed by Hanyang due to the economic agenda of the PRC to revive the industrial heritage of the areas.

Hankou's role as a commercial centre was interrupted after 1949.[5] Wuhan was chosen as the administrative centre of the Lianghu region because of Hubei's central location at the connection point between local waterways and adjacent provinces. In the following decades, Hankou's economic role was downplayed in local discourse, while China's Communist leadership adopted the Soviet model of economic planning. This state agenda began to change with the gradual implementation of national economic reforms as of 1978. In Wuhan's tri-city complex, Wuchang was the administrative and educational centre, whereas its industrial root would be traced to Hanyang. Hankou was once the local centre of railway transformation, but a new train station was built in Wuchang in the 1990s, and small stations have encircled Wuhan in all directions.[6]

Overall, Wuhan's status as a hub of industry and transportation was rising, while Hankou's status as a commercial centre was marginalized. As inter-provincial trade in grain came to a halt during the 1950s, food supply policies of self-reliance and grain-first production quotas became critical in agricultural sectors. With the large-scale expansion of urban-based industries, Wuhan ballooned beyond its earlier limits. After several rounds of administrative changes to its perimeters, it expanded from 302.7 square kilometres in 1950 to 8,212 square kilometres in 1985 (1,557 of which were urban).[7] The municipal districts of most cities in Hubei stabilized via internal administrative shuffling rather than incorporating adjacent counties. These administrative boundaries often shifted around Wuhan, indicating that it had established its position in the surrounding area.

Air Pollution in Wuhan

By the late 1970s, regular reports written by officials of the Hubei Provincial Bureau of Environmental Protection (HEP) regarding factory emissions were filled with growing concerns about airborne pollution.[8] Many HEP investigations frequently cited petitions launched by civilians in protest against air pollution. In December 1979, Pixian Chen, departing secretary of the Hubei Provincial Committee of the CCP, publicly addressed the seriousness of pollution in the province at a mass assembly. A few weeks later, the provincial administration issued a "Number 1" annual directive addressing public awareness of environmental protection.[9] A passage, from the same speech given by Chen in the above conference memo, recorded in December 1979, illustrates the official attitude to pollution:

> The task of environmental protection must be stressed, especially itemized into the [provincial] plan, while policy measures must be sincerely carried out. Units that have already installed pollution-control devices must not abandon such devices because of operational difficulties. Units that do not yet own such devices must accelerate the pace of their installation. Units that have caused serious pollution must be changed within a definite time frame. The central government has issued the Environmental Law. It must be strictly and thoroughly implemented; to do otherwise is illegal.[10]

Citing the reports of HEP officers, Chen made particular mention of certain cases of urban air pollution, including those in Hankou, Qingshan, and Huangshi, going back to the late 1960s.[11] Also in 1979, the central government invited three American environmental scientists to conduct a field survey of air pollution. Their June report recorded a worrying deterioration in the air quality of Beijing, Shanghai, Guangzhou, and Wuhan. All the scientists had a confirmed affiliation with the United States Environmental Protection Agency, yet a HEP internal update report revealed no personal details or speciality information regarding them. The HEP cited their survey as indisputable evidence of urban air pollution. Its abstract, initially circulated among provincial government branches, eventually reached lower-level offices via internal memos or formal directives.[12]

Earlier official responses to urban air pollution had blamed a lack of economic foresight. In a report filed in 1973, the Beijing government admitted making mistakes in its early design for manufacturing, which showed signs of "myopia" in industrial planning. Some HEP officers and their underling agents quickly responded that waste emissions issues should be raised and attended to by the national authorities.[13] In 1977, another report was circulated in Hubei, urging that Beijing should invest more in curbing pollution.[14] In Wuhan, official attention to air pollution caused by industrial emissions led to policy debates about allocating more investment in state-run plants.[15] The impacts of some environmental protection policies on local economies also evidently caught the attention of the HEP and its subordinate offices. Overall, administrative measures had already reached their limits in Hankou, Wuchang, and Qingshan.

Hankou

In the 1970s and 1980s, Hankou contained three of Wuhan's six industrial zones and most of its city-run industrial and commercial firms. In 1979, thirty-seven firms, operating mostly in Hankou, received administrative warnings that they could be shut down immediately due to their pollution.[16] The long blacklist included firms in the pharmaceutical, chemical, and textile sectors, plus five military logistics plants and five hotels. The warning of a shutdown sounded serious. However, making these polluters close their doors would be prohibitively expensive, and there was too much local employment and fiscal revenue at stake. Most of the companies were initially fined, and waste-processing fees were subsequently levied. However, the relevant documents reveal few figures for the fines or fees. Stressing "necessary" reimbursement, the fines were probably merely token amounts. Neither details of punishment nor the halting of operations are documented in the reports; nor are the names of individual protesters cited in the inquiry reports. Nonetheless, one can still surmise some official responses to urban protests against air pollution. Here are a few lines from the pollution investigation into the Number 2 Pharmaceutical Factory (also called Jiu'an, or "long safety") in Hankou:

> For years, the masses have repeatedly protested against the serious pollution caused by Wuhan's Number 2 Pharmaceutical Factory. Some protesters give the factory a nickname, "Inflictor Mill," and recite, "Jiu'an ['long safety'], Jiu'an, you are safe, [yet] we are not." Many residents, working for government branches, hospitals, schools, factories, and street districts, have petitioned officials of the municipal or provincial governments, even the central government. Some protesters posted *dazibao* [big-character political posters] on the factory front gates and directly scolded the factory managers in person.[17]

Although Hankou had a great deal of air pollution, this case received more public exposure than similar ones mentioned in directives and internal memos. In mid-1979, the Wuhan Bureau of Environmental Protection reinvestigated about twelve of the city-owned plants.[18] The continual discharge of industrial dust was linked to respiratory disease and other health concerns. The majority of urban air-pollution cases were in Hankou, but no valid solutions were found. A few reports proposed that plants engage in more internal recycling of industrial dust.[19] However, acts of voluntary control seemed far beyond the factories' capabilities, both in terms of investment and trained personnel. Factory managers and local offices in Wuhan, such as the Wuhan Bureau of Environmental Protection, repeatedly asked for investment that targeted air pollution.[20] By the mid-1980s, Wuhan appears to have prioritized the control of airborne industrial emissions, most obviously in Hankou.

Reports of air pollution in Hankou provide background information for more specific cases in the adjacent districts of Wuchang and Qingshan. A scattering of investment statistics appears in the early declassified pollution files. Compiled between 1985 and 1998, this urban dataset reveals that the annual total number of pollution-control projects jumped from 173 in 1985 to a peak of 494 in 1987; the level remained above 400 annually until 1990, declined to 313 in 1991 and gradually to 52 in 1997, but it climbed back to 132 in 1998.[21] Total annual investment peaked in 1991 and diminished thereafter. Among projects dealing with five kinds of pollution (industrial effluent, dust emissions, solid waste, noise, and others), investment for airborne pollution and effluent containment took turns occupying the top place in the pollution-control investment allocation of Wuhan in 1985–98.[22] Annual investment growth rates (based on the previous year), ranging from the lowest rate of 57.92 percent in 1986–87 to the highest rate of 62.48 percent in 1990–91, show that Wuhan expanded its pollution control activities significantly from the mid-1980s. More data are needed to confirm whether the decline in annual investment in pollution control during the late 1990s was caused by channelling into a smaller number of projects or reflected statistical standards adjustment. The situation is complicated by a lack of data on specific projects.

Unsurprisingly, economic returns on investment in waste control turned out to be mostly inadequate to sustain the air-pollution containment devices autonomously managed by factories. Alternative policy measures seemed necessary when increasing pressure came from urban residents, including both factory workers and government employees. This pressure would be articulated most in the form of letters of appeal, which were often paraphrased in HEP reports. For example, the reports frequently use phrases such as "the voices from the masses are rising every day" (*qunzhong husheng riyi gaozhang*).[23] Occasionally, to add more weight to their reports, officials opted for stronger phrasing, such as "the masses are extraordinarily indignant" (*qunzhong yichang jifeng*). Such short phrases seem to reflect sympathy for the protesters by at least some HEP officers, who dealt directly with petitions requiring immediate intervention.[24] Relaying urban protesters' demands for emissions containment, the officers played an intermediary role in negotiating conflicts of interest.

Wuchang

In Wuchang, two cases of air pollution stand out in the official documents, both of them closely associated with academia. The first example, in 1978, is from the Shizhishan (Lion Hill) district and involves the Huazhong University of Agriculture and the Hubei Academy of Agricultural Sciences. Both lie in the southern periphery of Wuchang. As a legacy of Zhidong Zhang, governor general of Huguang (Hubei and Hunan), both institutes originated from an experimental state-owned farm; influential alumni worked mostly at local agricultural institutions. In 1978, the university and the academy jointly filed a formal complaint in which they accused the Shizhishan Chemical Factory of continually emitting dust in their communities, as well as over the farmland and fisheries of the surrounding villages. They requested immediate mediation of the environmental protection offices.[25] However, neither the Hubei Provincial Archives nor the Wuhan Municipal Archives possess any follow-up reports regarding this issue, and it is doubtful that they were ever filed.

About eight years later, the second case pertaining to air pollution in Wuchang was filed by some "unnamed" staff and faculty members of the Zhongnan University of Law and Economics. Zhongnan's alumni worked mostly in local government and public organizations, including many state-owned plants.[26] Its neighbours included the Huazhong University of Agriculture, the Wuhan Textile University, and the South-Central University for Nationalities, all in the South Lake area.[27] Air pollution on

Zhongnan's old main campus was caused by emissions of fluoride dust from a factory owned by the Wuhan Institute of Chemical Engineering Research in the Qianjia Street district of Wuchang.

In August 1986, a public letter of appeal was jointly signed by the Zhongnan University of Law and Economics, with another prestigious educational institute, the Huashi No. 1 Middle High School, which was affiliated with the Huazhong Normal University. An abstract of the letter contained an "emotional" accusation, citing statistics of fatal diseases claimed to be related to factory emissions. The letter listed recent deaths, "evidently all directly caused by cancer," including a total of fifty-three staff members. Eleven had worked for the high school, and the remaining forty-two worked for the Zhongnan University of Law and Economics. According to the letter, "astonished" people witnessed that the tree leaves on the two campuses and nearby residential districts had turned yellow due to the factory's emissions, while many plants had withered or died. The letter was supported by two sets of investigative reports by the Wuhan Institute of Plant Science and the HEP monitoring station: fluoride residue found in pine trees on both campuses was about double the norm.[28] The monitoring report added further details:

> Since July 2 this year [1986], several faculty/staff members of the Zhongnan University of Finance [Law] and Economics have repeatedly appealed to the government agencies assumed to be responsible; yet our requests have not been appropriately addressed. Now some staff members have a more imperative appeal, protesting their right as citizens to public health. They have urgently demanded that government units with accountability strictly apply the environmental law to respond immediately to environmental pollution caused by the Wuhan Institute of Chemical Engineering Research, for the sake of quickly resolving an imperative issue of public health as reported by this large group of teaching staff members.[29]

This written appeal sounds noteworthy in the HEP reports, and the two cases of protest against air pollution have similar background elements. The reports for the second case cited more details. Only a few letters of appeal, such as the one above, were extensively quoted in the HEP reports; most of the paraphrased letters were not archived. Yet, nuanced responses to these semi-official opinions are still detectable. Compared to other cases, the 1986 petition from the faculty members resulted in a certain degree of success, probably because of the status of literati was rising during this early reform period in China. The factory accused of causing air pollution was ordered to suspend all operations. After installing equipment to control its emissions, it promised to scale back the output of its hazardous product, as soon as it delivered existing orders.

These two cases reveal the opinions of local literati in the public sphere.[30] In most cases, however, the HEP received information in the shape of what it formally called *gongmin jubao* or *shimin jubao* – "citizens' reports" or "city-residents' reports" and "anonymous tip-offs." The inquiry reports filed by local officials of the HEP typically reveal only minimal information about individual informants. Most official reports of pollution investigations also express approval to encourage the exposure of polluters. While environmental protection offices struggled with underfunding and understaffing, the cases detailed above also demonstrate how the HEP sought to gain all the support it could from other local organizations. Pollution incidents were filed with even greater urgency in the districts of high manufacturing density such as Qingshan, where locals demonstrated even more impatience and frustration.

Qingshan
Protest against air pollution peaked during the late 1970s in Qingshan. As was typical, the HEP reports rarely mention the names of victims or those who reported the problem, but in this regard, they probably followed bureaucratic protocol. The HEP officers' approach of indirectly citing protest was probably a routine part of the job of widely circulating information regarding pollution. Although the names of petitioners are typically not mentioned, the reports show that the HEP and its municipal officials cooperatively handled most cases. This pattern was manifested in an early air pollution case of 1979, when ninety-nine workers disseminated the contents of a public letter they had written. A July 1979 memo written by HEP officials stated,

> Recently [in June 1979], ninety-nine citizens from Qingshan, Wuhan, co-signed a public petition letter, requesting the investigation of serious pollution by the Wuhan Qingshan Sulphuric Acid Factory. We have accompanied

the directors of the Hubei Provincial Bureau of Petroleum and Chemical Products to conduct another joint field investigation and more negotiation, and we plan to take more committed measures to fully resolve factory emission problems.[31]

The letter from the ninety-nine workers expressed frustration over the limited measures that had been put in place to contain airborne emissions from the sulphuric acid factory, which belonged to the Wugang Group. The number of signatures on the letter is also worth noting, as ninety-nine sounds serious enough to get attention but is not large enough to suggest a mass protest. In China, the number nine implies "many," and thus two nines (as in ninety-nine) deliver a double sense of "many." The letter stated that the petitioners were planning to appeal to the Qingshan district court. Another public letter, probably from the same people, was addressed directly to Li Renzi, the first secretary of the Wuhan Municipal Committee of the CCP. Claiming that they would organize strikes in schools and factories, a few protesters attempted to enter the factory, which had already been locked to avoid head-on conflicts.

Municipal and provincial authorities quickly responded to calm the protest. The HEP stepped up its communication with municipal districts and local industrial bureaus to de-escalate the situation. Funding then arrived to finance dust-control projects at the factory, including some new emissions-collecting devices and a taller chimney. Predictably, the HEP follow-up reports regarding this episode omit details of strikes, yet they confirm the financial assistance for the equipment upgrades.[32] Production at the factory was suspended for closed-door inquiries. However, the Wugang Group's overall production was not interrupted.[33]

In addition to factory workers, students, engineers, and teachers at the Wuhan Institutes of Iron and Steel signed the letters, playing an active role in bringing air-pollution problems to public attention. According to a HEP internal circular, some students and staff informed a group of foreign engineers, who had been invited to Wuhan to provide technical training and who also took photographs as evidence of rampant air pollution. While local city-district government officers warned that this would undoubtedly create a seriously negative image among outsiders, the HEP quickly issued a temporary sanction to halt factory operations.

Few further details were filed, but this case received much attention before the HEP closed the case. Its officers invited senior officers from the Provincial Bureau of Petroleum and Chemical Products to an onsite investigation and talk about the governance difficulties they themselves faced. Spontaneous acts of exposing pollution indicated rising environmental awareness. The protest letters were also an increasingly regularized vehicle for local environmental protection offices to gain public support and justification. Choosing this protest method probably allowed anxious workers/residents to procure more resources from local and central authorities: air pollution was most visible in industrial zones, which were usually adjacent to residential districts.

Compared to Hankou and Wuchang, Qingshan was a site of newer heavy industry, and the solutions to fighting its air-pollution problems seem to have been more technologically oriented. Dealing with dust emissions appeared to be framed in terms of technological upgrading and cost-benefit considerations. All the proposed solutions required continual investment. However, the existing equipment and any additional resources for dealing with emissions created little immediate benefit but required more follow-up investment and expensive maintenance. Moreover, each of the above industrial pollution cases can be seen in a cynical light: once a plant was officially labelled as a polluter, its manager could reasonably expect that its investment budget for technological upgrades would be expanded. One can imagine the sense of frustration and futility experienced by local environmental protection officials because these state-owned plants and their supervisors would be reluctant to fully enforce environmental regulations, fearing the effect on production and revenue flows.

Soil Pollution in Rural Hubei

Hubei's urban environmental protection officers tried to defuse public protest against air pollution in the city. Does the same story also apply to soil pollution in the countryside? The answer is "yes." County environmental protection officials worried about how to relay rural concerns about industrial pollution, usually on a small scale, that were voiced in informal letters of appeal addressed to decision makers. Despite frequent complaints to their superiors about their meagre resources, some officers made impressive efforts to ensure that the concerns of rural residents reached higher-ranking officers and national policy makers.

In 1974, the topic of urban pollution became the focus of the first policy report issued by the National Leadership Office of Environmental Protection, which proposed priority areas for environmental protection policy. The critical domains summarized in this report included soil contamination and food security, directly relating to rural welfare.[34] Local government offices came to realize the level of seriousness resulting from the contamination of farmland, as reflected in directives from higher-level offices.[35] County environmental protection officers also largely played the role of arbitrators between people who complained about the pollution and those who created it, who typically professed innocence or admitted to only partial responsibility. Administrative protocols for soil pollution cases resembled those for urban air pollution: first, the offenders would be warned that they might be mandated to halt production, the victims of pollution would receive financial reparation, and finally, a token cleanup of the affected land would be undertaken.

The two cases of soil pollution in Daye and Dawu counties are presented below to reveal the efforts made by county environmental protection officers. Many HEP reports refer to soil contamination in rural areas that was caused by toxic chemicals or heavy metals. Many inquiry reports clearly state that those accused of dumping effluent or solid waste were perfectly aware of the consequences of their actions. Some simply hoped that they would not get caught. Less noticeable than air pollution, soil pollution affected a more isolated area, while creating a long-term impact with more irreversible damage to rural communities. Some local voices brought it to public attention. The soil pollution cases in Daye and Dawu bluntly reveal conflicts of economic interest, particularly between the manufacturing and agricultural sectors.

Daye County, the Residue of Early Industrial Experiments
Daye county is proud of its mining tradition, but its iron ore reserves were eventually exhausted, and its industrial glory had faded by the late 1970s. The rich history of the Daye Mining Company can be traced to the Hanyeping Company, discussed in Chapter 2. During China's large-scale

industrial experiments in the late nineteenth century, the Daye Mining Company was Hanyeping's iron ore supply base. Bordered by the ancient city of Ezhou, the configuration of the county's administrative area had remained largely unchanged. However, recurrent changes in its administrative status imply a complicated story. At various points after 1949, Daye was apportioned to Wuchang and four surrounding prefecture-level cities: Xiaogan, Xianlin, Huanggang, and Huangshi. This arbitrarily assigned county-city status was confusing, and Daye was reinstated as a city in 1994.[36] In China, county status usually implies an economic and administrative focus on agricultural production. Urban status normally indicates more of an emphasis on industrial output, generally measured more in quantitative rather than qualitative terms dating back to the 1970s and 1980s.

In Daye, economic conflicts emerged between industrial sectors and rural areas. Villagers suffered from soil pollution, at roughly the same time that some early directives were issued by provincial authorities to maintain fisheries, and Daye county authorities became increasingly concerned about industrial pollution.[37] The mining sector generated massive quantities of waste, and the iron and steel mills also created large amounts of debris. The city of Huangshi reimbursed fishing and farming villages in Daye for damage to their fisheries and crops caused by mining and manufacturing operations.[38] The governor of Hubei, Han Ningfu, appointed in 1980, was shocked by the gravity of pollution in Daye: roughly 40 percent of its rice production was contaminated by a variety of heavy metals, particularly cadmium.[39] In revealing this alarming evidence, the HEP reports explicitly compared the situation to that of Japan in the 1960s, where eating rice grown in cadmium-laden soil near mining sites had resulted in a series of tragic cases of Itai-itai disease, which causes often-fatal renal dysfunction.[40]

A HEP memo for a provincial conference shows that in 1980 Daye county officers openly apologized for failing to protect the land. Though apparently stirred by reports from officials at lower levels, the governor, as conference chairman, did not seem prepared to accept the shocking statistics. His doubts about the accuracy of the statistical analysis were documented. At an internal meeting, when he wondered whether some figures might be subject to speculation or fabrication, he was immediately refuted by Daye county officials sitting in the back rows, who bluntly referred to a set of early analytical results made by the hydro-ecology experts

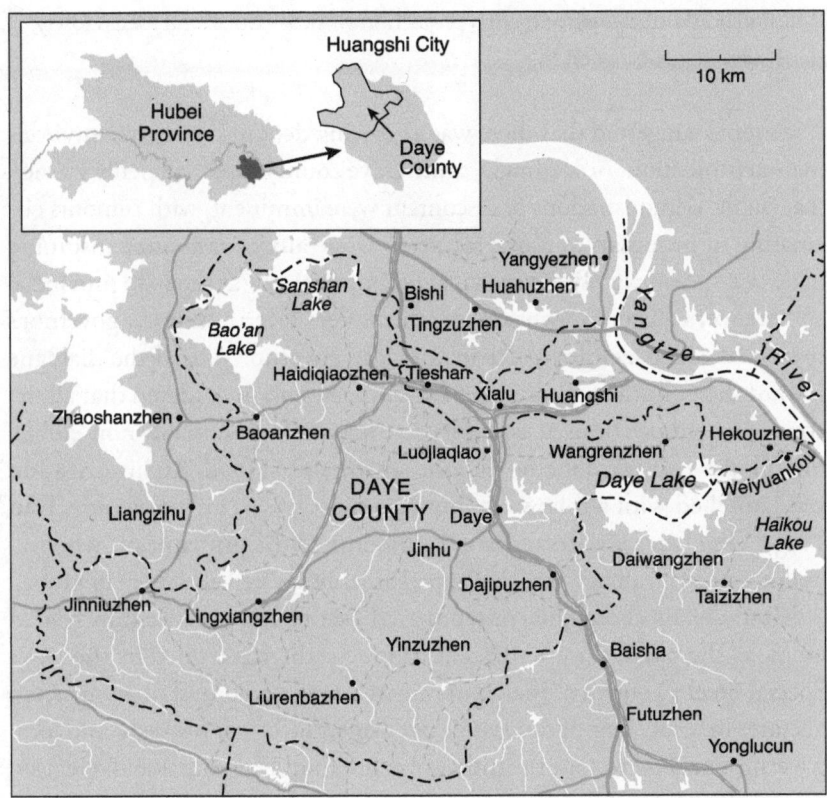

FIGURE 8 Map of Daye, Huangshi, Hubei | Hubei Provincial Platform for Common GeoSpatial Information Services. *Adapted by Eric Leinberger*

of HEP monitoring stations in Wuhan and Huangshi. Some county officers continued to complain that superintendent officers did little to help, simply demanding more paperwork. Requests for more intervention were relayed to Huangshi. Another series of directives cascaded down within the bureaucratic hierarchy.[41] A 1979 directive (in the form of a memo) stated,

> Over the past several years, Hubei has done much work to regulate industrial waste. However, in some areas, industrial pollution remains serious. The masses have strongly appealed. There will be no change to stressing the environmental protection and pollution control goals within the national economy adjustment agenda. All the CCP committees and administrative

units should treat environmental protection work as critical and focus closely on carrying it out well.[42]

The memo suggested that there was a growing demand for better environmental protection. Accordingly, some Daye county officials openly warned that public demonstrations of discontent were imminent, with rumours circulating of proposals to block factories, while anxieties about consuming toxic rice spread widely among villages. Impressively, at the 1980 provincial conference, a few unknown county officers kept interrupting the governor's speech to the assembled participants. The governor shifted the dialogue toward the pollution solution for Daye. In the end, he ordered that all the contaminated rice be used as factory raw material (presumably non-edible, though the associated memo did not say so), not as food. An investigation was launched with a follow-up inquiry report by the vice-governor, Tian Ying.[43] The Daye reports repeatedly underlined the principle of "whoever pollutes must clean up," which was presumably applied elsewhere in Hubei.

More details about the contaminated rice paddies surfaced in related memos. The paddies were adjacent to sewage ditches owned by the Daye Special Steel Company. The contaminated area amounted to about 1,320 hectares in total. Tens of thousands of villagers lost their harvests, and local governments had to foot the bill to procure food from outside the region. Senior officials harshly condemned the crisis, including Li Xiannian (1909–92), a Hubei native who had recently become vice-chairman of the CCP. Li demanded, "Shut down the polluting plants!" However, he added the qualification "if all else fails."[44] Nothing failed as seriously as the Daye case. Surface soil pollution was deliberately diluted by irrigation water with abundant seasonal rainfall. Villagers struggled to acquire safe drinking water, and the county paid for the construction of new water treatment facilities and delivered freshwater by tanker trucks. Expressing sympathy while mortified about their powerlessness, the Daye officers persistently complained to higher offices, and the HEP circulated scores of stricter measures to prevent further incidents.[45]

Dawu County, Home of Rebellious People
Located within the prefectural jurisdiction of Xiaogan, Dawu enjoys a renowned status stemming from its revolutionary tradition, a legacy celebrated in local history.[46] According to official narratives of the CCP,

hundreds of thousands of Dawu folks joined the Red Army from the 1920s to the 1940s and were killed either in internecine wars against the Chinese Nationalists and other warlords or in the Second Sino-Japanese War (1937–45). Dozens of Dawu natives were well-known generals in the People's Liberation Army and/or were senior party leaders, who designated Dawu as a cradle of the revolution.

Dawu is a hilly county. A substantial portion of its arable land comprises hillside terraces containing thin layers of soil. Unlike rice-producing Daye, it grows corn, yams, potatoes, peanuts, and tobacco. It is frequently afflicted by floods, landslides, drought, hail, wind, and plagues of locusts. Industrial activity was almost statistically negligible in Dawu before the electrical grid was extended into it during the late 1960s, when rich deposits of iron and copper ore, as well as phosphorus, were discovered in Dawu. This event opened a new chapter, one both blessed and cursed.

Dawu had long struggled with self-sufficiency and poverty before intense conflicts emerged during the 1970s between mining interests and nearby villages. Largely isolated by mountainous terrain though it was, Dawu was not immune to soil pollution. Its first official record of soil pollution caused by industrial waste emissions dates from 1970.[47] In the mid-1980s, industrial waste reached a peak, far exceeding the capacity of nature to absorb it. In 1981, the county government swiftly approved the establishment of its own environmental protection office along with a large group of pollution monitoring stations across the remote hilly landscape on the northern border, apparently in response to the serious results of soil contamination. However, in 1987, sixty chemical-products-related plants in Dawu were reported still to be emitting waste, with little proper treatment. A short entry in the *Dawu County Gazetteer* noted that twenty thousand mu of agricultural land were officially recognized as being seriously damaged, coincidentally the same figure as reported for Daye just a few years earlier. The matching amounts of damaged cropland make a meaningful connection between the two counties.

Ironically, the discovery of the rich phosphorus and iron ore deposits, which helped support the resource-depleted iron mills of Daye, had fuelled hopes for the improvement of Dawu's social welfare and economy. Investment in the mines had entailed a special policy for the county, as a poor area that was also a cradle of the revolution. However, its well-known but scarcely documented case of soil pollution is typical in that it illustrates

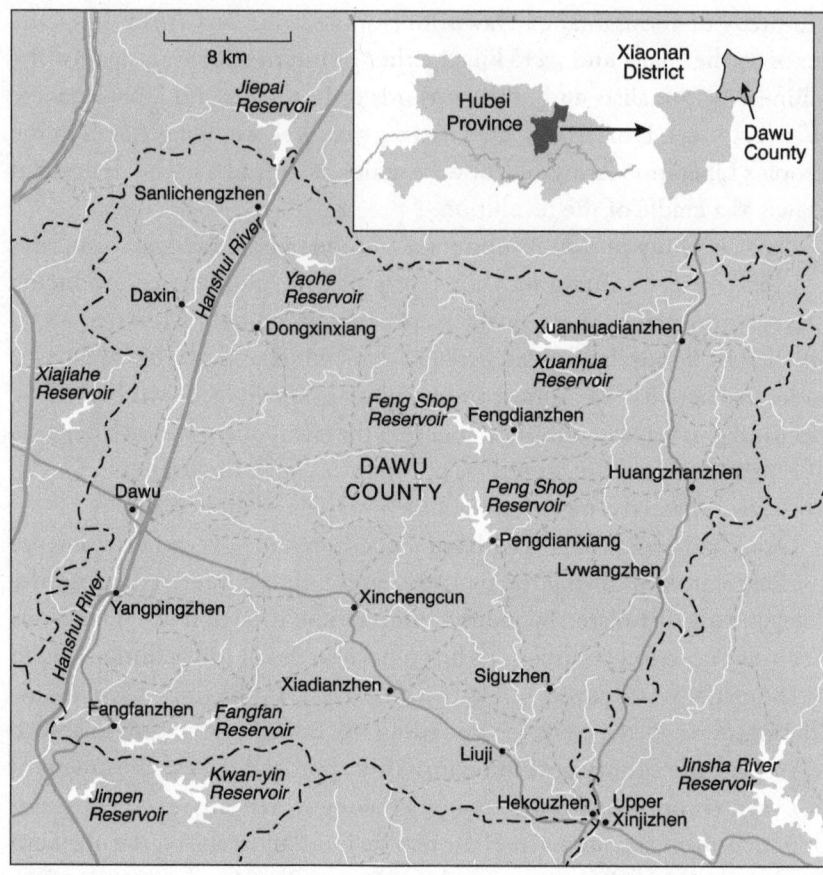

FIGURE 9 Map of Dawu, Hubei, which is now part of Xiaogan City (known as Xiaogan Prefecture before 1993). | Hubei Provincial Platform for Common GeoSpatial Information Services. *Adapted by Eric Leinberger*

the massive environmental costs caused by the negative spillovers of large-scale resource extraction. An investigative report on industrial pollution correlated to mining activities from 1979 described outbursts of popular anger, first noted by sympathetic lower-level officials:

> Because pollution problems have not been effectively resolved for so long, public health has been severely impaired. Many local environmental protection office representatives continually report that … the masses have some strong resentment toward industrial pollution. Written petitions from the

masses fly in like snowflakes. Some officers have still not paid due attention. The environmental protection task is not on their agenda; many issues cannot just receive a service ticket number and then stand in line for replies. There is no time to read these reports.[48]

Mining persisted in Dawu into the 1990s, when its phosphate reserves were finally exhausted.[49] Chemical fertilizers and pesticides, particularly DDT, one of the most important chemical products manufactured in Dawu, were widely used in the 1970s, and this increased crop yields. However, in that decade, few were aware of the cumulative damage that pesticides and fertilizers caused to ecosystems. In 1983, the use of DDT was banned in China after its long-term harms were recognized.

Because the mining sector generated revenue and provided employment, local authorities had to support its growth. But agriculture was also important, a conflict that reached alarming proportions when agrarian losses due to contaminated soil far exceeded the revenues produced by mining. Quicker to recognize the situation than county officials, some villagers sent letters of appeal to relatives who now worked either in the central or provincial governments. Moral support was helpful, and most appeals received sympathetic replies. Yet county authorities could not neutrally resolve the competing interests of peasants and factories who shared the same space. Industrial interest was vested in the state-owned mills that, though situated physically within the county's jurisdiction, were legally beyond it. Even the highly respected generals of Dawu who had survived the battles and internal conflicts of the early revolutionary years could not freely express their views on industrial pollution.[50]

Educational Reforms as Intellectual Legacies of Wuhan

Regarding the question of how the intellectual legacy of Wuchang connects with cases of industrial pollution, it is possible that the central government provided more effective air-pollution management in Wuchang simply because the intellectuals who lived there could exert significant influence in terms of relevant policy consulting, making and implementing. Some may seem more capable of advocating for anti-pollution measures because of their connection to the Wuchang intellectual establishment. The attention paid to, and the priorities that stem from, these types of policies are generally found in communities that are better educated, have more influence, or who have more advantages than their less-informed counterparts. Residents of such communities are able to use their knowledge or status to achieve higher living standards and a better quality of environment for themselves, as they have more sway with Beijing in connection to anti-pollution measures. Here is my main argument to the investigations above concerning air pollution: protests voiced by Hubei's civil societies are connected to Wuhan's cultural and educational legacies. Wuhan has the third-highest number of college students in China, after Beijing and Guangzhou. According to records from 2019, its registered college students had reached a stunning figure of 1.126 million, nearly half of which, 0.527 million, were women, in ninety colleges and universities; most of their campuses were in Wuchang.

Wuhan's 1985 yearbook indicates that most local schools, ranging from elementary schools to colleges, started at Wuchang and then spread to surrounding areas. Since 1949, school regrouping and openings of new schools have attempted to balance school-district distribution with the expansion of urban, suburban, new industrial, and residential sections. The school system penetrated Wuhan's urban and suburban areas, yet most colleges and universities still clustered in Wuchang. A straightforward reason exists for this: from the Ming to the Qing period, it was the seat of the provincial government. In addition, men who wished to become imperial officers sat their *keju* exam in Wuchang, at the Phoenix Hill Examination Hall, which lay beside the provincial offices in the designated administration zone. Dating from the Su Dynasty (581–618 CE), the keju was a gruelling Confucian exam for the selection of civil servants; it was discontinued in 1904 during the last dynasty. Until the late Qing, candidates for the civil service in both Hubei and Hunan wrote their exams in the Phoenix Hill hall of Wuchang.

From the Ming Dynasty (1368–1644) to the Qing Dynasty (1644–1911), the three old towns of Wuhan had twenty-one academies, twelve of which were in Wuchang; two of the most famous were the Jingxin Academy and the Lianghu Academy, whose main campus the Wuhan Conservatory of Music has inherited today. Some of Hubei's first modern Westernized schools emerged in Wuchang. One of the earliest private universities in

China, Zhonghua University, was founded there in 1912; Wenhua College, also in Wuchang, was created in 1871 with funding from the American Episcopal Church. In 1950, Wuhan had twelve registered universities, eight of which were in Wuchang. In 1985, twenty-eight of thirty-four universities in Wuhan had their start in Wuchang. Campuses are typically near small hills or slopes or urban lakes. Most universities prided themselves on the serenity of their campuses; locals also used the names of the hills as a form of shorthand in referring to the campuses, namely Luojia Hill for Wuhan University, Yujia Hill for the Huazhong University of Science, Guizi Hill for Huazhong Normal University, Lion Hill for the Huazhong University of Agriculture, Snake Hill for Zhongnan University of Law and Economics, Shipai Ridge for the Wuhan University of Technology, and Nanwang Hill for the China University of Geosciences (Wuhan). All of these nationally known academic institutes are scattered around Wuchang.

Wuchang played a prominent role in the educational reforms of the late 1970s. The reforms helped popularize some dynamic adaptations while simultaneously creating dissent, invoking the need for conflict intermediation but also leading the way toward market-oriented reforms under the orders of Deng Xiaoping in 1977–78.[51] Two significant men, Quanxin Zha and Daoyu Liu, both connected to scholar groups in Wuchang, played a legendary role in the educational reforms initiated by the CCP. A highly respected professor of electrochemistry at Wuhan University, Quanxin Zha (1925–2019) headed its Chemistry Department in 1980–84 and was selected as a member of the Chinese Academy of Science in 1980.[52] This membership can represent the highest respect and privilege paid to elite Chinese scientists.

Anecdotal accounts in memoirs and news reports state that during an internal professor-scholar senior-consultant symposium chaired by Deng Xiaoping in July 1977, Zha rose to his feet to deliver his concern and interrupted the proceedings. He abruptly warned about the declining quality of university students, which he linked to the abolition of the national college-entry examination. It had been discontinued during the devastating chaos of the Cultural Revolution (1966–76), when "worker-peasant-soldier" students were admitted to universities via the so-called mass recommendation. The lack of a screening exam left the door open for irregularities and corruption in student admissions. Other participants at the meeting immediately confirmed Zha's words. In response, Deng simply asked for suggestions as to when the entry exam should be reinstated. Zha answered, "The sooner, the better. Not one single day to waste."[53] In October, Deng responded: a national directive announced that provisional college-entry exams would begin in December 1977. Without most of the previously imposed restrictions, the exam was open to all high school graduates and youth students who had been sent to villages; it was regularly held every June-July thereafter. Without exaggeration, the lives of millions of Chinese citizens altered their trajectories thanks to this exam.

Daoyu Liu (1933–) also attended the 1977 meeting at which Zha voiced his criticisms. A reputable alumnus of Wuhan University, he had helped organize the event as director of the Higher Education Department at the Ministry of Education.[54] President of Wuhan University from 1981 to 1988, he was one of the youngest university presidents in China.[55] After undertaking pioneering reforms at the university level, Liu was removed from this high-profile position, after gaining an enormous reputation with some radical criticism when he was just fifty-five years old. The authority at the time, likely the Ministry of Education, disclosed little reason for his removal, except for citing a long-term health issue that Liu himself also mentioned in explaining his transfer from Beijing, as a senior officer in the Ministry of Education of the PRC, to his newly appointed post as the President of Wuhan University, Hubei, his home province. Liu's reforms were both resisted and endorsed. His "radical" experimental measures included the reintroduction of the Western college course-credit and professor-tutor system, releasing course registration restrictions, and promoting competition among faculty and staff. In later years during his unintended retirement, Liu seems to have become a less ardent reformer. Yet he kept publicly commenting on education corruption and injustice. Nonetheless, after he stepped down from his post in 1988, he never held another public position that attracted remotely as much attention as his last one as the President of Wuhan University.

From the late 1970s, China's campaign to reform education can be roughly divided into three progressive stages: Education Modernization in 1978, Education Industrialization in the mid-1990s, and Education Internalization, whose agenda overlapped with that of the second stage. From 1975 to 1979, as increasingly publicized in state propaganda, the "Four Modernizations" were re-introduced and applied to agriculture, manufacturing, science and technology, and national defence. Education was categorized under science and technology, a seemingly sensible arrangement. Local leftists still criticize the exam "reform" that Deng reinstated in October 1977 as being a counter-revolutionary and "backward" step: the *keju* exam system was used for centuries to select imperial bureaucrats from Confucian students. Many CCP cadets were discontent, and they argued that Deng's college-entrance exam acted like the keju exam, a policy tool for sustaining the bureaucratic scholar-official governance. Today, most local and central education policy makers in China often celebrate their role in restoring the cultural tradition of civil exams, which serve to verify skills and the grounds for policy formation.

Before 1999, university funding in China came mainly from the central and local governments. The market power changed the rules of resource distribution after the educational system started adapting to market-oriented reforms. Less burdened by a state-dominated pattern, Wuhan was in a better position than many other cities of the PRC to procure educational investment. Controversially, rumours spread that the proposal for an "Industrialization of Education" came from a renowned economist, Dr. Ming Tang, who wrote directly to the central committee of the CCP in November 1998.[56] Tang had earned his PhD in economics from the University of Illinois at Urbana-Champaign in 1989 and had spent some years studying mathematics in Wuhan. In 1998, he was the chief economist in the Beijing office of the Asian Development Bank.[57] Acting on Tang's advice, the higher education sector expanded swiftly with "industrialization" after 1999, deemed as the start of a controversial enlargement of university student recruitment and infrastructure investment, which has persisted to the present day. A 2015 survey noted that, in Wuhan, the ratio of college students to urban residents was 137 to 1,000.[58]

Trade-Offs

Weighing up the trade-offs between economic growth and environmental protection has added complexities to both rural and urban spaces in China.[59] As its administrative system lacked clear or formally recognized avenues for shaping environmental protection policy governance, letters of complaint, like the ones discussed above, became the main form of expressing discontent. The state responded to them with varying degrees of receptivity.[60]

Protest seemed tolerable when it was not voiced by a large group of people and still remained under state control. Hubei's archives shed light on how certain societal pressures helped push environmental governance within the local bureaucracy. Both continuities and ruptures have evolved in China's environmental history from ancient times.[61] Local government authorities have also played a pivotal role in mediating and implementing the balance between economic institutions and communities. In various regions of China, environmental management issues have also been closely associated with some popular support levels for government policies by local civil societies and the legitimacy of local and central rulers.[62] Although the present inquiry does not extend beyond Hubei, we can see that many officers worked assiduously for better environmental governance, even though local administrative systems were often blamed for obstacles to environmental protection or the effective regulation of pollution.[63]

As mentioned, city-dwellers more likely bore the brunt of air pollution, while rural people were more directly exposed to soil pollution.[64] The authorities in Hubei tended to pay closer attention to protests from urbanites than to those from villagers, and contemporary reports were dominated by urban complaints regarding air pollution. It created obvious damage, whereas cases of soil pollution were comparatively isolated and more limited in scope. The localities and identities of pollution victims underlay sad disparities between urban and rural spaces; reparations to pollution victims differed among socio-economic groups, which fuelled resentment and created challenges for inter-group negotiations.

Hubei's environmental protection officers seem to have agreed among themselves that the consequences of industrial pollution differed for the agricultural and manufacturing sectors. Although pollution in both urban and rural areas continually threatened people's livelihoods, manufacturing interests received priority over those of the countryside. Local HEP officials were committed to containing pollution, but their regulatory practices had minimal impact. At best, their work can be described as falling short of expectations; at worst, it can be seen as a total failure by today's standards. In their defence, it should be noted that they had neither the willingness nor the capacity to override state policies, which emphasized industrial growth. They knew perfectly well who would be reading their reports and what kind of response they could expect. They did act on some interventionist calls for environmental governance.

In many routine reports on pollution, negotiation can be seen between local HEP offices, petitioners, and polluters. This "negotiated symbiosis" implies that the Chinese state would not necessarily always be dictating top-down edicts to local societies, and the nature of such state-society symbiotic relations would thus empower China's burgeoning civil society with a diffused or informal government-relation network.[65] From time to time, environmental protection officers did attempt to mediate between competing interests. However, most of their measures did not generate sufficient positive results or alter the fundamental fact that the growth of the state-owned manufacturing sector would always take precedence.

Conclusion

For years, communities in Hubei struggled against industrially generated pollution in the air and the soil, evoking public health issues and conflicting interests between various sectors. Environmental protection officers felt compelled to toughen anti-pollution measures with better governance. Reading between the lines of their reports, one can also apply the *sine ira et studio* (without fury or bitterness and bias or partiality) approach to better understand how they took a large portion of the blame for their failure to achieve a great deal. They were usually described as a crowd of faceless and nameless actors (both male and female). Nonetheless, they were members of the community, and their socio-economic identity was much like that of other urban and rural residents. Recognizing that industrial production must be maintained and that state-assigned output quotas also must be met, they followed the policy mandates of Beijing. Economic means for dealing with pollution, including processing fees, waste-discharge fees, and penalties for violating regulations, were supported by administrative protocols and commonly employed at the provincial level and below. Ultimately, they had little lasting impact, as incidents of pollution frequently recurred in the same spaces, and communities were obliged to call on the environmental protection officers yet again. During the early reform years of the 1970s and 1980s, the officers continually filed follow-up reports, often with sympathetic but seemingly useless remarks.

4
Struggles for Policy Implementation: Establishing the Environmental Agencies

In many of its image promotion pamphlets, Wuhan brands itself as a city of *jianghu*.¹ The word jianghu literally means rivers and lakes, but in China's popular culture it can also refer to an imaginary alternative world. Often rendered as a secular, enigmatic, and clandestine underworld or culture, jianghu is informally presented in some public discourses. It is characteristically associated with tumultuous waters that symbolize the prospects for perilous odds and blessings, paradigmatically leading to either remorse or heroism in nostalgic and popular memories. In jianghu stories, various forms of rebellion and resilience are eternal themes. Thus, jianghu would not be typically endorsed in official narratives, and it would be even restricted or absent from local government offices' bureaucratic writings. Given this, an office established by a government body would seem to represent the diametric opposite of this alternative universe. Interestingly, however, many state officers in China frequently use "jianghu" to describe their work and their lives as public servants, though rather colloquially or informally. Like small boats in turbulent water, they see themselves as constantly subject to an unpredictable maelstrom of risks and uncertainties. This chapter will explore how government environmental protection agencies managed to find their footing in shifting waters.

In the hope of contributing to the knowledge of environmental policy-making in China, especially concerning state policy during the Mao era, this chapter traces the establishment of environmental state agencies at the provincial and subprovincial levels in Hubei from 1972. Of course, rampant pollution was not a new phenomenon in the province, which had documented a series of ecological crises as far back as 1957. In discussing China's environmental protection and pollution issues, many current policy researchers concentrate largely on national contexts rather than subnational ones. I recommend that we take the opposite approach and highlight the evolving roles of regional agents and policy continuity in the subnational structure.[2] Many scholars suggest that post-1949 China has made little progress in environmental protection. For example, Elizabeth Economy examines the constraints on efforts to preserve natural environments by containing industrial pollution after years of rapid economic growth. However, such a period of dramatic evolution, though seemingly full of ruptures, still exhibited continuities within the local contexts.[3] Few scholars inquire into the establishment of the subnational environmental agencies who directly engaged with practical issues. Of those who do, many focus on negative factors, including a lack of transparency, manipulation by elites, and bureaucratic weakness.[4]

The roles of provincial environmental protection agencies deserve more specific inquiry. Hubei's archival records reveal that a group of offices took responsibility for pollution abatement much earlier than previously thought, even though China's era of mass movements (*qunzhong yundong*) crimped most of their efforts.[5] Their emergence should be credited mainly to the "state campaigns" of the Mao era. Moreover, I would suggest that subnational administrations did not have the same policy priorities throughout China.[6] These findings support the sympathetic approach by David Allen Pietz, whose inquiries focus on the marginalized state agencies of Maoist times.[7] Historians also need to describe how local endeavours responded to national mandates, particularly regarding how the sophisticated policy-making methodology of "proceeding from point to surface" (*youdian daomian*) evolved with the features of policy experiments and "model experiences" in China, as examined by Sebastian Heilmann.[8] Along with China's socialist industrialization after 1949, the tenacious effort to "revolutionize" its agricultural production galvanized the efforts of many agriculturalists to farm scientifically.[9] Nonetheless, the numerous pollution reports in Hubei underscore that provincial officers, along with their municipal and county equivalents, played a pivotal role in regulating pollution abatement. However, most studies tend generally to agree that China's central state agencies played a dominant role in environmental pollution affairs and that its regional offices had a limited role.[10]

Two mass movements are of particular relevance here. One is the Three-Waste Governance Movement (Sanfei zhili yundong), which targeted urban industrial pollution. The other is the Biogas Promotion Movement, whose purpose was to expand energy experiments throughout the countryside. In both movements, a series of factors hampered early environmental protection efforts, including inter-agency bureaucratic conflict, underfunding, and the difficulty of effectively regulating state-owned enterprises. The biogas movement did not begin as an environmental concern but may have later developed into one. Traces of proactive practices for environmental protection emerged in Hubei from 1957 in its early experiments with a biogas economy. Compared to the Sanfei zhili campaign, the biogas campaign seems more successful than is recognized in the current era of chemical fertilizers. Early efforts to enforce more strict environmental protection faded after the earlier reform era of the late 1990s, which arguably started to disincentivize the state-sponsored collective projects in post-Mao China. Both movements are seemingly absorbed into other local agendas of environmental governance, while no official statements that could have provided closure were ever released.

It is necessary to define what is meant by a state "environmental agency": Staffed by government-designated personnel, it is funded by the government and must therefore adhere to state protocols. It handles various regulatory issues, and its daily chores extend from policy design to implementation, including regulating waste emission, promoting pollutant-abatement measures, and applying any policies and codes with concerns about ecological impacts. In Hubei, there were four such groups.[11] Of key importance was the Hubei Provincial Bureau of Environmental Protection (HEP) and its affiliated facilities, officially established in 1979. Most provincial environmental offices were founded earlier than the HEP, but they operated less independently and remained mostly within established frameworks.

The second agency is the Hubei Provincial Leadership Office of Biogas Construction (or the Hubei Biogas Office for short), established around 1975 and also known today as the Hubei Provincial Rural Energy Office, which is subordinate to the Provincial Ministry of Agriculture.[12] Some may argue that it was not an environmental agency, as it was part of a biogas-promoting state campaign whose purpose was to resolve the energy deficit in rural areas. In response, I would argue that rural energy consumption played a crucial role in energy conservation and deforestation. No matter how doubtful it may sound, given the Maoist regime's general record of ecological degradation, the biogas offices deserve credit for diversifying energy sources.

Vaclav Smil advocates the use of small coal mines, small hydro-stations, and household biogas as a viable (and probably upgradable) approach to China's rural energy predicament.[13] Rarely examined, the biogas case challenges the prevailing assumptions or critical views on China's environmental protection policies. For instance, in 2001, Judith Shapiro examined environmental degradation in Maoist China, relating it to four core themes: political repression, utopian urgency, dogmatic uniformity, and state-ordered population migration, which was a prevailing critical assumption at the time.[14] In 2012, she highlighted the roles of China's subnational administrations in enforcing public policy goals or initiatives for environmental protection and showed how governments and the public responded differently to local environmental crises with relevant regulation challenges.

Two additional groups of environmental agencies had seemingly much less independent funding or personnel support than the HEP and the Hubei Biogas Office. This book refers to the third group as a few powerful offices in the provincial executive branches, including the Provincial Planning Committee and the Provincial Health Department. For instance, this group of planning committees functioned within the framework of the Planning Committee from the national level to the provincial, prefecture, and county level; it reflected some need to shift red tape among subdivisions of already well-established government branches. The fourth group, including a phalanx of offices in the provincial industrial bureaus, consisted of their waste-emission control offices. Their overseers either merged with or shrank into industrial offices or associations of metallurgy, machinery, textiles, mining, construction material, electricity, and petroleum-chemical products. The fourth group gradually vanished during a set of market reforms in the 1980s.

The Hubei Provincial Bureau of Environmental Protection

Many records indicate that ecological deterioration in Hubei caught the attention of state authorities as early as 1970, in the midst of the Cultural Revolution, perhaps China's most dramatic period of chaos. The Cultural Revolution spanned ten years, but its high tides clustered around its beginning in 1966 and its end in 1976, though calamitous aftershocks certainly lingered. During this period, industrial pollution raised public concern for living environments that were deteriorating with natural resource depletion. Many studies concentrate on Beijing in interpreting the evolution of China's environmental protection policy, yet subnational governments did enact their own policy to some extent. Either openly, semi-publicly, or internally, subnational officers wrote many reports on local ecological crises.[15]

The 1970s saw a series of national mass movements or state campaigns, all involved with pollution regulation in one way or another. In 1974, a rudimentary version of the HEP evolved in two parts. The first was a small leadership office whose purpose was to oversee and coordinate industrial pollution policies.[16] The second consisted of a network of environmental protection monitoring stations (EPMS), with state-funded onsite or field research facilities. The leadership office was replaced by the Provincial Leadership Office of Environmental Protection in 1975, which became the HEP in 1979. Most scholars mark 1979 as the beginning of the "open-door policy" in China, after decades of exclusion in the Cold War. It is also the year that – for the first time – Beijing passed a provisional version of its National Environmental Protection Law, which has been continually reviewed and modified ever since.[17]

Part of the origin story of the HEP, Hubei's EPMS network's preliminary functioning dates from July 1974, though its operational facilities had long existed under the watch of the provincial administration of public health.[18] These EPM stations appeared to be a tangible result of the Sanfei zhili Movement traced back at least to 1970 when they were officially established and documented in later records, deputizing the provincial pollution monitoring system aimed to contain emissions of three industrial wastes (*gongye sanfei*): wastewater, dust-fog, and solid waste.[19] The Hubei provincial Sanfei Office, a predecessor of the HEP, had assumed the task of regulating industrial waste for years, while the public health authority had also shared the responsibility for monitoring pollution, particularly of air and water. The National Leadership Office of Three Industrial Wastes Governance was founded to materialize policy resolutions for industrial pollution even earlier than those Sanfei offices in Hubei. Its establishment was a response to the increasing frequency of regional pollution reports, a trend that would suggest frustration at policy effects and public pressures regarding incidents of rampant pollution.

Administrative warnings were frequently used but seemingly seldom effective, largely thanks to an institutional deficiency of accountable state-agencies of environmental protection, particularly in their early stages after being established. The HEP was founded on a provisional basis, preceded by the Hubei Provincial Leadership Office of Three Industrial Waste Control (or the Sanfei Office for short) and the Provincial Leadership Office of Environmental Protection, with its EPMS network. As revealed by a series of emergency notices and internal reports, early signs of declining quality in water bodies received negative attention in "the province of lakes." The human-geography features of Hubei exposed its populations to effluent pollution. The provincial administration centre was situated at Dong Lake, where, presumably, staff could see that it was seriously polluted by industrial effluent. In the spring of 1970, the Hubei Provincial Revolutionary Committee issued an urgent warning to subordinate committees that water pollution in Hubei had attracted the attention of Beijing.[20] During the early establishment of the Sanfei Office in 1970, some CCP traditions came into play: opening an "office," rather than a committee, board, commission, department, ministry, or bureau, indicates that the venture was short term or had an ad hoc agenda. Overseen by the Hubei Provincial Construction Committee, the Sanfei Office would view its work as a provisional task of policy coordination among other executive branches of the provincial government.

An emergency directive of 1970 shows that the Hubei provincial administration was dissatisfied with the efficacy of subprovincial officers in controlling industrial pollution, which was caused mainly by massive machinery-manufacturing industries in urban areas. It recommended stricter inspection and waste abatement measures. These effluent pollution "accidents" affected rural areas the most and raised serious concerns regarding public health. A 1972 report by the Hubei Provincial Bureau of Water

Utility and Irrigation revealed that water pollution was widespread and severe.[21] To this report was attached a backlogged supplementary 1979 laboratory report filed by the Wuhan Municipal Disease Prevention and Monitoring Station (DPMS), documenting the health consequences of effluent pollution in Ya'er Lake.[22] Every year, more than 200,000 tons of fish were caught in the lake and were consumed mostly by urbanites in Wuhan. As incidents of eating toxic fish repeatedly occurred, the spread of effluent pollution caused the death of all the fish in the surrounding fisheries. This case depicts how subnational negotiations of pollution issues resulted from nascent concerns about ecological decline, as observed firsthand by provincial and county officers.

In connection with Ya'er Lake, the county government of Erchen pointed the finger of blame at a chemical factory in the town of Gedian, which was under the jurisdiction of Wuhan. In fact, Erchen had been complaining about the factory since 1966, alleging that it was pouring chemical waste into the water bodies connected directly to the nearby lakes. Requests for immediate intervention were filed away, while the unregulated emission of poisonous water continued for years, with the damaged fisheries as a background. At the beginning of the Cultural Revolution, no county authorities would risk investing their meagre resources in this less prioritized issue, despite worrisome signs of ubiquitous ecological decline.

As revealed by later reports, the emergency directive of 1970 had little effect in reversing environmental deterioration. Official responses all came out of health concerns caused by unregulated or under-regulated emissions of industrial effluent. In fact, local grievances about emissions dated from even earlier, as shown in a 1964 memo issued jointly by the State Planning Commission, the National Department of Health, and the National Bureau of Metallurgy.[23] All industrial waste emissions raised public-health concerns, particularly in connection with the iron-steel sector. The Provincial Department of Health filed an emergency notice in 1968, addressing health issues caused by industrial waste.[24] Warning of the spread of chronic diseases, the notice updated the cases of "occupational-hazard" illnesses associated with a list of manufacturing sectors, including mining and metallurgy. The recipients of this notice included an extended list of industrial bureaus, a few of the largest general hospitals, and the DPMS network. The EPMS system slowly came out of the existing DPMS network, accompanying the elusive and loosely recorded formal establishment of the Provincial Leadership Office of Environmental Protection.[25] Monitoring industrial waste emissions started as "sideline" work by these research units hosted in the provincial and municipal public health stations.

More reports poured in, calling for immediate action. In May 1973, the Hubei Sanfei Office filed a directive and relayed it to subprovincial governments, as well as state-owned factories and schools.[26] It provided an update on pollution in Dong Lake and projected growing discontent. In August 5–20, 1973, Beijing held the first national environmental protection conference, as a follow-up report by the State Planning Commission informed the provincial governments in 1974.[27] Notably, the conference proposed launching national and regional offices for environmental protection and the control of industrial pollution.[28] The Hubei Construction Committee relayed the national report to its sub-branches and other concerned administrative branches. Following this inaugural national conference, Hubei convened its own provincial environmental protection conference, on October 9–13, 1973, which was held in Wuhan.[29] It appeared to have little immediate impact, as more incident reports were generated and warnings regarding effluent issues continued to flood in.[30]

Nearly eleven months after the provincial conference, the Hubei Sanfei Office finally proposed the official forming of the Hubei Provincial Leadership Office of Environmental Protection in 1974.[31] The proposal was accepted and then turned into the formalized establishment of the latter roughly around 1975. A vice-director of the Hubei Construction Committee was designated as its head director, with a staff of twelve officers. The new office reported to the National Leadership Office of Environmental Protection in Beijing. In January 1974, the Hubei Sanfei Office issued a mandate to launch the EPMS network, which would be affiliated with the DPMS network. A memo was filed to establish this new HEP office, with a separate budget for the EPMS network, to the branches of the central government, including the National Ministry of Health and the National Leadership Office of Environmental Protection in the State Planning Commission.[32] The Hubei Construction Committee approved the EPMS budget, forwarding the memo to the DPMS (Disease Prevention and Monitoring Station) at both provincial and municipal levels and a few health research institutes that specialized in occupational diseases for some mining and metallurgical sectors in Hubei.

In 1976, the EPMS network started to consider splitting away from the DPMS system.³³ The two networks shared the same staff and facility resources in Hubei's health administration. Inter-agency and Intra-agency power struggles ensued: the superintending department filed requests for more budget solutions and intermediation. In 1977, the EPMS network merged with two municipal research institutes at Wuhan and Huangshi, gradually forming a single entity yet still operating locally.³⁴ The new institute became known as the Hubei Research Institute of Environmental Protection. This move initiated a formal split between the DPMS and EPMS systems. A total of 150 officers (70 for Hubei, 50 for Wuhan, and 30 for Huangshi) eventually worked under the Hubei Provincial Leadership Office of Environmental Protection. The Sanfei Office transferred all its former duties to that office. The latter became the HEP in 1979, without even updating its official seals in some reports before a provisional version of the first National Environmental Protection Law passed that same year.³⁵ The newly established HEP finally had an independent budget for thirty-five officers as a starting point.

The Hubei Biogas Office

This section will investigate the institutional evolutions of the Hubei Biogas Office, which is also known as the Provincial Office of Rural Energy.³⁶ Such a comparison may seem irrelevant, yet the evolution of the two entities helps to tease out more interpretation of local context. Both agencies shared some early features, after which they evolved in different directions. The Hubei Biogas Office, also a provisional provincial body, started with ad hoc policy missions and was grounded in a mass movement that began in 1975. It arose directly from the promotion of the biogas economy in rural China. One can interpret this alternative solution for rural energy and fertilizer production as a response to the growing problem of organic pollution, mostly in soil and water, by untreated human waste and animal manure. Hubei's biogas economy deserves some attention, and its biogas office heralded subsequent rural energy policy experiments in China.

A quick review of relevant studies provides some context for the launch of the biogas movement. Writing in 1984, Vaclav Smil synthesizes cultural, political, economic, and historical analysis, and notes the dire state of

China's rural energy and its ecological decline. He calls for a decentralized supply of rural energy and recommend that biogas technology would pose a viable solution to China's rural energy shortage.[37] Smil's work draws mostly on secondary sources using official reports, while he tends not to address regional cases. Pessimistic about the biogas economy, he comments that China's complex bureaucracies, insufficient or uncoordinated, are good at "promulgating new laws and regulations and holding grand conferences (disguised banqueting mostly)" but generally "less adept at getting things done."[38] This unflattering opinion is echoed in some of the documents in the archives of Hubei.

However, some researchers take a more positive stance. Overseas observers often refer to Sichuan as an archetypal case of China's biogas legacy, illustrating how, from 1975, its provincial governments helped to propel a national mass movement to promote biogas. In 1972, Sichuan began to host a series of national colloquiums about the use of rural biogas; provincial delegates came from all over China. A manual for constructing a water-pressure biogas digester was widely circulated in about 1976, encouraging more experimentation with digester-construction materials and feedstock to improve designs that were suited to regional conditions.[39] This crash-course program provided very practical how-to lessons to rural trainees, who subsequently returned to their prefectures as teacher-technicians.[40]

From Sichuan to Hubei, we can detect a nuanced transition of state policy implementation regarding the evolutionary paths of the CCP's "model experiments," from dispatching cadre working teams to reliance on local cadres in the post-Mao era.[41] This promotional pattern in the local biogas economies also featured "adaptive R&D and incremental innovations," a demonstration model credited most often to Sichuan for potential use by other provinces, as indicated by some inquiries on Hubei's early biogas-promotion policy experiments by Moulik.[42] The biogas movement was based on the rural collective structure of commune, brigade, or team, and the number of digesters gradually expanded, resulting in a coverage of 70 percent of rural households as "basically" popularized by 1979.[43] By 1975, Sichuan had more than 400,000 digesters; by 1979, it had 5 million.

Sichuan alone accounted for more than half of the biogas projects in China in 1979. Nationally, the Chinese people constructed 7 million rural household digesters in 1973–78, in addition to 30,000 large-scale systems serving communes, hospitals, schools, sewage plants, distilleries, and other facilities that produced organic waste. However, from then onward, the total steadily declined, dropping to around 4 million by 1983.[44] This reversal resulted from more digester units abandoned than new ones built. Its leading cause was a response to adjustments in national policy.[45] Subnational responses indicate lingering concerns about state policies, partly attributed to post-1979 reforms. Another explanation plausibly points to the introduction of the agricultural responsibility system in 1978.

Reforms replaced the mode of collective agricultural production with an approach dominated by individual household leaseholds. Changing incentives shrunk large-scale rural public construction projects. Also, the drop in the number of digesters resulted from technical problems, including poor design, alternative construction measures, and maintenance issues. Deficiencies in policy adjustment can result from institutional constraints, as Smil implies. Referring to some policy shifts, he advises that the best outlook for China would be to pursue "some gradual localized improvements and the prevention of further major degradation in key sectors and areas."[46] The mass movement of "popularizing" biogas would enrich revisionist narratives regarding China's environmental legacies if we agree that Chinese rulers gained a sense of legitimacy and popularity by assisting the people in environmental crises.[47] The biogas case thus helps highlight that the Chinese state is capable of responding to local environmental management issues with mass mobilization. One may argue that a golden rule or universal standard of legitimacy, such as democracy, should be accepted across cultures and national borders, yet it looks like differing views may also make sense.

Like Sichuan, Hubei played a remarkable role in promoting the biogas economy, though it has received less notice in Western scholarship. Hubei's local government authorities showed great interest in promoting experiments with the biogas economy as early as 1957. Many biogas records are scattered throughout the branches of the provincial government; all of them lead to a report by a scientific research unit based in Wuhan.[48] Subtitled "Supplementing Gasoline, Diesel, and Coal," this report on bioenergy utilization research included policy proposals and technical instructions while calling for more trials; it documented the details of experiments with biogas devices, adding some interesting first-hand illustrations and diagrams. The unusual way in which this policy proposal was submitted is noteworthy: In January 1957, two officers working in the Zhongnan Materials Research Institute wrote a public letter to the Hubei Provincial Revolutionary Committee of the CCP. It included some detailed proposals for promoting the biogas economy in the province.

This "People's Letter" (Renmin laixin) was endorsed by provincial authorities. A list of state-funded research facilities and universities in Wuhan, the Chinese Academy of Science, Wuhan Branch, and Wuhan University, co-signed the letter. The Hubei Provincial Bureau of Industries quickly approved the report and delivered the funding for more equipment via its laboratory. More local research institutes, such as the Huazhong Agricultural Research Institute and the Wuhan Bureau of Machinery, joined a set of biogas field experiments in April-May 1957. In June 1957, the team successfully ignited a methane lighting system; in August, it achieved satisfactory preliminarily results for a prototype digester that used cow manure as raw material and produced biogas for generating electricity. The assessment report gave credit to Soviet experts who had lent their unpublished research notes, though a key device that measured biogas saturation came from a Japanese institute.[49] Encouraged by these promising results, another research unit, the Bio-energy Utilization Research Team, quickly began operation under the supervision of the Hubei Provincial Bureau of Industries in September 1957. The biogas device was tested as a replacement for straw, wood, coal, and oil as rural energy or renewable fuel for canteen lighting.

An additional economic-benefit report was generated, along with a technical analysis report.[50] After this, the archival records are silent regarding the fate of the pilot project, a lack of information that probably correlates with a period of political chaos that swept China. From 1957 to 1959, Beijing pursued a nationwide Anti-rightist Campaign and launched the Great Leap Forward Movement around late 1958, which culminated in three years of famine and ended in 1962. In 1958, coinciding with a series of catastrophic effects in the Great Leap Forward, more articles were published that promoted many technologies related to local self-reliance. In 1960, another letter revitalized biogas experiments in Hubei; it was written to the provincial government by Biwu Dong (1886–1975), former vice–prime minister, future vice-chairman and acting chairman of China.[51] An internal meeting quickly updated new experiments with more recent, spontaneously conducted cases; barriers in labour and construction materials were further assessed; focus shifted to the shortcoming of fuel depletion; the conclusion specifically underlined the benefits of preserving fuel resources, adding manure supply, and improving sanitary conditions.[52]

The vice-chairman endorsed biogas technology with great interest. His second letter to provincial CCP leaders urged them to invest more resources in the biogas economy. Referring to the progress of neighbouring Anhui province, Dong advised that all digesters, some of which had already been abandoned in Hubei, should be restored in stages. In May 1960, Hubei organized a two-week conference to exchange county lessons for popularizing biogas and another (non-fossil-based fuel) energy-saving technology, the so-called industrial microwave.[53] The latter turned out to be a hoax, but the biogas policies survived and were popularized nationally around 1975.

Soon in 1960, the Hubei Provincial Committee of Science and Technology circulated a three-year policy proposal for industrializing biogas, underlining that it helped reduce energy costs, conserve natural resources, especially coal and wood, and promote sanitation plus other benefits. In August 1960, it was replaced by a seemingly more realistic report, which called for a five-year period and further experiments.[54] This less-aggressive agenda provided flexibility to subprovincial agents, and state-owned farms among the earliest practitioners of biogas knowledge.[55] During the 1980s, Beijing relinquished control of some state farms, which then came under the jurisdiction of a county or a township. These local agricultural syndicates had assumed many public functions, including public health, hazards watching, elementary education, and community security.

The popularization of biogas occupied part of the core agenda of most state farms, while county governments played a coordinating role. Hubei's biogas practices seem less significant than those of Sichuan, but archival records show that Hubei probably took a more flexible and less politicized path. This change implied more localized discrete policy implementation for the promotion of the biogas economy; local governments could invest more efforts in prioritizing their policy agendas. In 1978–79, the Hubei People's Revolutionary Committee filed an administrative directive; the biogas office started first with four officers in 1978, expanding to eighteen staff officers in 1979 and growing steadily from then on.[56] Witnessing the benefits of the biogas economy, Hubei's county agents organized training programs following the example of Sichuan. Representatives of rural communities visited successful sites before deciding to invest; then more villagers registered for training and certification to earn extra income as technicians to superintend construction of biogas devices.

Hubei gradually formed a biogas service station network. Its biogas economy can be accredited first to the Provincial Leadership Office of Biogas Utilization, currently known also as the biogas office in Hubei, a predecessor of the Hubei Provincial Office of Rural Energy. The network of offices began to produce policy reports for improving rural energy consumption. By 1983–84, Hubei boasted 60 county biogas service companies with 427 employees (up from 342 in 1979) and 127 biogas service stations with 1,365 workers.[57] Despite the long-term efforts of these biogas agencies, historians rarely examine their contribution to the improvement of rural energy. Hubei's early experiments with a biogas economy suggest that its subnational offices willingly advocated an alternative vehicle for a recycling economy as early as 1957.

Inter-Office Evolution

Both the HEP and the Hubei Biogas Office sprang from mass movements around 1975 on an ad hoc basis. The Sanfei movement targeted industrial waste emissions in cities, whereas the biogas movement addressed energy problems and fertilizer shortages in the countryside. The biogas movement left traces of spontaneous initiatives in Hubei. Remarkably, when the biogas movement was initiated in 1975, Chairman Mao received full credit for it. Briefly visiting Wuhan on April 11, 1958, to view an early biogas device, he had written a short comment: "This [biogas] must be well promoted" (Zhe yao haohaode tuiguang). Seventeen years later, his remark was interpreted as vital guidance for the launch of the movement.[58] The fact of his visit was rediscovered only recently, and neither Biwu Dong's letter of 1960 nor the biogas officers of Hubei mentioned it. In 1975, the Beijing government elevated the biogas movement to the level of the Learn from Dazhai in Agriculture Movement (Nongye xue dazai). As part of its promotion, three national agencies (the State Planning Commission, the Chinese Academy of Science, and the National Department of Agriculture and Forestry) co-hosted the first national conference in the same year to encourage experiments with a biogas economy.

In the 1960s and 1970s, the Provincial Committee of Science took charge of stimulating the biogas economy that had finished most of its early experiments. Based on the 1960 letter by Biwu Dong, the National Department of Agriculture, however, was in charge of promoting the biogas economy. In 1979, the Hubei Biogas Office was regrouped under that department, like the Bureau of Land and other internal offices.[59] A recurring pattern of cyclical decline had partly led to policy failure for this sustainable economy. In February 1996, the Hubei Biogas Office became the Hubei Provincial Office of Rural Energy, but it retained its former title for external use. In August 2007, the provincial government issued an administrative recommendation that all sub-provincial biogas offices should operate jointly with the offices and bureaus of rural energy and eco-energy. For the sake of inter-agency efficiency, the registration of two brands for either internal or external use with the same staff is a typical arrangement in China. County offices and bureaus of rural (or eco-) energy agencies gave some space to subregional agendas for rural energy, a policy inherited from the early years of the biogas economy. Soon, a similar agency arrangement also applied to the Hubei Provincial Institute of Environmental Monitoring and the EPMS network affiliated with the HEP.[60]

The intent of the joint-agency strategy may have been to achieve bureaucratic efficiency, but it also suggests a compromise arrangement in the allocation of funding and personnel. Many pollution reports confirm inter-agency collaboration; when investigating pollution in the vast rural regions, the city-based HEP agents relied heavily on the agricultural agencies, because villagers constituted the primary victims of industrial waste emissions. Cooperation among offices supported policy continuity; newly established offices also needed to maintain a good connection with their former superiors or peer institutions. This arrangement demanded an alignment of interests. Competition for administrative resources partially mirrored this fragmented bureaucracy, despite the nominal promotion of inter-agency assistance in many reports. For the HEP and its subordinate offices, the insufficiency of deployable resources hampered policy tasks. Still, many government branches played a role in regulating pollution, and inter-agency investigations seem to have been common. In 1979, for instance, the internal office of science and education in the Hubei Provincial Department of Agriculture conducted an overview report on rural ecological deterioration as the principal investigator, designated by the National Department of Agriculture.[61]

Influential provincial executive branches included the authorities of planning, health, and construction, and all of them retained internal offices to coordinate pollution issues and specific sectoral regulations. Among these influential branches, two deserve closer scrutiny: the Provincial Planning Committee and the Provincial Bureau ("Department or Ministry" then, now "Commission") of Public Health. Differences between a bureau (*ju*) and a department or ministry (*ting*) usually correlate with degrees of status in the civil service hierarchy and affect budget priorities and levels of responsibility.[62] Most internal environmental protection offices vanished along with their supervisory agencies; the provincial bureaus of textile industries, chemical industries, and mining industries all regrouped into a provincial industrial association during the reforms of state-owned enterprises. The HEP and the Hubei Biogas Office were the survivors, benefiting from the reforms.

The benefits gained by the HEP look more notable than those enjoyed by other offices. The HEP began its life as an office, became a bureau, and finally evolved into a department. Initially, it was under the Hubei Construction Committee, but today it operates with increasing priorities in provincial policy agendas. By comparison, the Hubei Biogas Office has always remained a subordinate entity while inheriting the tasks from the Hubei Provincial Bureau of Industries, with its 1957–58 experiments, and then the Provincial Committee of Science in the 1960s and 1970s. Afterward, the Hubei Provincial Department of Agriculture oversaw its long-term mission. In a series of ongoing reforms, the Provincial Department of Forestry split off from the agriculture department; the Provincial Bureau of Fisheries, the Provincial Bureau of Animal Husbandry and Veterinary Service, and the Provincial Agrarian-Ecological Protection Station all remained under the full or partial supervision of the Hubei Provincial Department of Agriculture.[63] The Hubei Biogas Office has maintained a status equal to these bureaus, and all of them are tasked with specific duties for rural ecological protection, further typifying how China's fragmented bureaucracy evolved with environmental policies.

When Abigail Jahiel (1998) examined the national structure of environmental protection agencies in China, it was noted that many of their local offices had succumbed to ongoing reforms.[64] Another article by Dieter Grunow (2011) also examines subnational state agencies whose missions involved environmental protection and links them with the four groups mentioned at the beginning of this chapter.[65] Some provincial agencies were once actively involved in waste emission regulation, but now only a few documents hint at their existence. This vanished group, mostly affiliated with industrial bureaus, often interacted with a dominant government branch, the Provincial Planning Committee. Responsible for assessing funding requests by state-owned industries, this committee played a key role in representing its superintending agency, the State Planning Commission, which was a "mini-cabinet" in the central government. At the subprovincial level, city or county planning committees authorized new projects to coordinate the centrally controlled projects. In this process, factories would file applications for pollutant-abatement funding, and their emission-control projects would then need to be registered as budget items, either as continuing investment or new construction investment.[66]

The Sanfei Office of the Hubei Provincial Bureau of Metallurgy is an example that helps to demonstrate how these internal offices were marginalized.[67] As mentioned earlier, Hubei hosted several industrialization experiments during the late nineteenth century, one of which was in the town of Hanyang. Part of Wuhan since 1949, it was the site of China's first modern iron facility: the Hanyeping Company. During the war against Japan (1937–45), its assets, including its equipment and staff, were sent westward to the wartime capital of Chongqing. In the 1950s, assets were transferred to two industrial syndicates in Hubei; most went to the Wuhan Iron and Steel Company on the north shore of Dong Lake, and the rest went to the Daye Special Steel Company. The latter answered to the Hubei Provincial Bureau of Metallurgy whereas the Wuhan Iron and Steel Company operated directly under the central government. Ostensibly, the Sanfei officers in Hubei could regulate the waste emissions produced by the Wuhan Iron and Steel Company, but this was virtually impossible given that the company was owned by Beijing.[68] In 1979, it was included in a national warning list regarding industrial waste emission that issued a probationary period of three years (1978–81). The company was not obligated to respond to provincial or subprovincial bureaus, whose administrative rank was lower than that of Beijing. As a result, the Sanfei officers had no power to punish its misconduct.

Most Sanfei Offices in provincial bureaus served as regional deputies for their national agencies. For instance, the Daye Special Steel Company constituted the most substantial responsibility of the Sanfei Office of the Hubei Provincial Bureau of Metallurgy.[69] In 1972, the Hubei government issued several administrative orders, following the directive of the central government, to establish such Sanfei Offices across industrial bureaus.[70] The bureaus all held a double identity in the state-led economy: they were regulatory agencies, as well as industry representatives vested with sectoral interests. In 1979, the HEP increased its interaction with Hubei's metallurgical sector.[71] The Hubei Provincial Bureau of Metallurgy and the metallurgical sector received some positive assessments from the HEP for their cooperative work; the HEP rewarded the Daye Company and its affiliated mills, the largest province-owned industrial syndicate, for its progress in controlling emissions.[72] On the other hand, its assessment of the provincial chemical industry was less positive, and it recommended a fine plus a halt to operations as punishment.[73] Moreover, funding applications had posed a long-term issue for these state-owned enterprises and industrial bureau officers, dating back to 1972.[74]

Hubei's pollution reports give no indication that funding improved. Instead, they show that provincial and county governments increasingly resorted to economic measures to contain industrial pollution, including imposing waste-processing fees, emission-allowing fees, and fines.[75] Economic punishment became a modus operandi, as China started transitioning to a mixed economy in a series of market-oriented reforms from 1979. The formal legitimization of administrative punishment, as institutionalized by the National Administrative Litigation Law, was not officially approved until 1989. In that year, the HEP started resorting to court action to achieve its ends. However, Hubei provincial court records from 1992 to 2005 show that the results were not inspiring.[76] As part of this transition of environmental protection policy implementation, local governments paid more attention to economic growth, while subnational environmental policy challenged lower-level regulation officers for achieving pollution-abatement targets. Economic punishments were frequently criticized as mere slaps on the wrist, a shortcoming that Hubei's provincial and subprovincial offices admitted from time to time. Industrial regulators were criticized openly, but the enforcement of environmental protection and the governance of industrial waste emission remained mostly disappointing.

Provincial, municipal, and county environmental protection officers are sometimes charged with intentional or inadvertent negligence, a suggestion that is challenged by my findings. Most of these individuals served as witnesses and solution-providers in a context of ecological crisis. Residents of Hubei, they spent their lives in their communities and shared similar values with their families and friends. Lacking any proof to the contrary, accusing them of making political trade-offs for personal gain seems arbitrary and even occasionally unjustified, though some scandals probably did occur, attracting public attention. Therefore, if we wish to assess their performance in a neutral manner, we need to highlight their multiple identities and diverse experiences, and to take identity politics into account. The early cases of Hubei help reveal the issues that challenged environmental protection offices. Rather than dissecting various pollution cases under a microscope, I would stress that the offices faced probably one of the worst periods of ecological degradation. Nonetheless, they were among the first to react to the ubiquitous signals of environmental deterioration, and there is rich evidence for both passive and proactive responses in their reports.

Here is one caveat for further interpreting the archival evidence: I would neither exaggerate nor trivialize the linking role of subnational agencies in examining how China materialized relevant policy practices in the 1970s and 1980s. Regarding how institutional factors critically interacted with environmental policy implementation at the subnational level, internal discourse about conservation policies in Hubei can be traced to 1957, much earlier than the Cultural Revolution, which began in 1966. It was always more bureaucratically expedient to propagandize political agendas than to follow through on a policy commitment. The primary sources show that concerned voices in Hubei resonated with specific policy measures. In some instances, the frustration of officers is evident in their reports. The bitter failures of the early environmental protection practices and efforts at regulating industrial waste emissions can be attributed to many causes, while bureaucratic weakness is a crucial factor to explain the dire status of compromised regulation for containing industrial pollution.

Hubei's investigative reports on incidents of industrial pollution indicate that environmental protection officers never hesitated to do their jobs, resourcefully improvising from the vague national policies for containing industrial pollution. In return, policy failures of these early regulation events have indicated a series of bureaucratic weaknesses and continual regulation failures in those early reforms and establishment stages of local environmental management of Hubei, particularly in Wuhan. It was not their fault that the bureaucracy was fragmented and weak or that many early reforms were largely unworkable. Archival documents indicate that the provincial environmental protection offices, such as the ones in Hubei, had recognized the pressing ecological decline observed in the chaotic stages of the party-led mass movements of the 1960s.

Evidence indicates that Hubei responded to pollution issues early and received guidance from Beijing afterward. Its environmental protection offices unavoidably engaged with two national campaigns of remarkable significance: the Sanfei zhili Movement and the Biogas Promotion Movement. Signs of their efforts to improve the ambient quality of air and water would still be detectable, from reading between the lines in archival accounts. Despite the turmoil of the 1960s and 1970s, Hubei's provincial officers made impressive efforts to implement environmental protection policies. The archival records trace the long path they followed as they struggled to acquire funding and staff resources.

Conclusion

From the 1970s to the mid-1980s, the environmental protection agencies of Hubei were plagued by the bureaucratic weaknesses that typically afflicted subnational governments in China. Initially, numerous government branches shared regulation duties, but many did not survive, and ultimately a shorter list of offices, chief among them the HEP and the Hubei Biogas Office, were designated explicitly to administer environmental policy. In a period marred by political chaos, where funding and staffing were insufficient and the national industrialization agenda commonly trumped other concerns, these local environmental protection officers still made valiant efforts to do their job. By today's standards, the outcome was a dismal failure. However, our policy reflections ought to address more voices and perspectives to adopt a neutral stance; thus, when assessing accountability for this lapse, we must always remember to highlight local initiatives in the implementation of environmental protection policy.

5

The Right to Pollute: Resorting to Cost-Benefit Calculations

In Hubei, poorly regulated emissions of industrial waste had obviously caused most environmental problems. Although household wastes were also identified as a source of pollution, official efforts focused on the industrial sector. After the HEP was established in 1979 as a separate bureau, its agents led environmental protection efforts in the province. Failures in implementing environmental policy in both Hubei and China itself were part of the growing awareness of industrial pollution problems and the global need for regulation. The rapid expansion of industrial production in many places worldwide often damaged natural environments. Throughout most of these industrial production processes, particularly in their final stages, unregulated waste emissions provoked tensions between those who benefited from the polluting industries and those who suffered from the externalized production costs. As environmental policies aimed mostly to defuse or reduce such tensions, the chances of achieving a complete solution seemed remote. Economic conflicts were commonplace; for example, if it is to flourish, agriculture requires untainted soil and clean water, but manufacturing interests can damage both.

Many new social conflicts became salient in Hubei and demanded official attention, arising from clashing economic interests among farming, fishing, and manufacturing sites that were state-owned but locally operated.[1] The degradation of water and soil, apparent since the 1960s, threatened livelihoods and public health in rural areas. At the same time, the polluting industries continued to benefit from subsidies and other preferential state policies. Political instability exacerbated the situation, but even at the best of times, ecological and economic-social damages are difficult to assess. Economic calculations can only approximate the accountability of responsibility for waste emission. When people do not bear the full costs of their behaviour, they tend to be less careful in avoiding damaging behaviours than otherwise. Hubei's environmental agents still relied on cost accounting used in the state planning system despite the difficulty of calculating costs. This approach would imply trade-offs between long-term interests and some short-term benefits of economic activities within an overarching policy framework prioritizing the calculation of fiscal expenditures and economic benefits.

Among a small group of China's elite economists, one outstanding individual is Dr. Peigang Zhang (1913–2011), the celebrated founder of China's Development Economics. Notwithstanding the fact that Zhang marginally explored the topic of industrial pollution in his early research, his work has had a huge impact on the examinations of China's industrializing evolutions and rural-urban transitions. He was from Hong'an, a well-known county of "poor mountains and evil waters" in northeast Hubei, the home of 223 Red Army generals who were decorated in 1955. After graduating from Harvard University, Zhang returned to China in 1946 and chaired the Department of Economics at Wuhan University. He won the David A. Wells Prize in Economics in 1946–47 for his degree-thesis "Agriculture and Industrialization." During the Cultural Revolution, Zhang was denounced as a "reactionary academic authority" and forced to perform hard labour. After 1976, his academic privileges were restored, as a founding professor of the Huazhong University of Science and an honorary dean of the School of Economics. In 1992, the school established the Zhang Peigang Development Economics Prize in his honour, one of the highest awards to Chinese economists. Even though most of his influential studies were completed before the 1970s, his death invoked waves of posthumous honours.

During the 1970s and 1980s, a small group of Chinese economists addressed policy solutions for industrial pollution. I will use archival evidence to examine how the environmental protection agencies in Hubei enforced pollution regulations by imposing costs on manufacturing enterprises that came under their jurisdiction. Pollution abatement implementation outcomes can be partly interpreted as the logical and rational result of balancing policy negotiation with conflict resolution on economic behaviours. Although the concept of sustainable economic activity was unknown in the 1970s and 1980s, its basic principles, such as trade-offs, economies of scale, and cost-benefit analysis, were part of the common language shared by the elite Chinese economists and state officials, and were frequently used in pollution reports. As local environmental offices created no material products or generated immediate returns, their agendas relied on public investments. These long-term industrial investments or reinvestments gradually affected both producers and consumers.

The evidence indicates that from the 1970s, a series of sectoral codes concerning the containment of industrial waste emissions was gradually elevated to a strategic level in the national policy, which emphasized economic development. Prioritizing economic growth had been the norm after 1978, and more specific industrial waste emission codes were proposed. This entailed the installation of numerous city-based environmental protection monitoring stations. They were quickly built in Hubei, but the network needed more administrative resources and legal authority before it could function effectively. The preliminary results of HEP monitoring work fell below the expectations of citizens and the goals of Beijing. Hampered by limited resources, the HEP agents admitted that everything was new to them, and their early work schedule was occupied primarily by personnel training and device procurement. Their tasks included raising public awareness and popularizing concepts about clean production in local communities. To be fair, the HEP was essentially a consulting service. It was not effective at regulating industrial waste emissions. Yet even its consulting work seemed to fall short in the early stages. It was disappointing, annoying, frustrating, and occasionally controversial, and its weak regulator role marginalized the HEP.

Once again, my discussion highlights how the main players (referring first to local economists) narrated and conceptualized local development paths with both ruptures in governance as well as continuities in policy. In shaping the paths of decision-making and policy reactions, they helped produce an anthropogenic world that was characterized by both good and bad exchanges between humans and nature and between humans themselves. From a holistic outlook, a good governance system must include the maximum number of stakeholders. Institutional constraints can be generally summarized as insufficiencies in policy motives and implementation, industrial codes, and enforcement of regulations. Various conflicts of interest impede the achievement of national or subnational environmental governance goals. Although some scholars have addressed the subject of China's environmental law in passing, the archival records show that the process of legal reform during the post-Mao years helped legitimize environmental policies and the establishment of environmental state-agencies at all levels. In Hubei, the policy efforts to regulate industrial pollution engaged many stakeholders, yet the weakness of its bureaucratic hierarchy significantly limited the results.

Economists and Environmental Challenges after the 1970s

One can map the origins and the shifts in environmental-economic thoughts of economists in China's policy pursuit of a one-dimensional economic development agenda into a more balanced approach that mandates the protection of natural resources. My overview of academic discussions will take a detour to explore the trajectories of relevant economic policies in the PRC. Particularly, its most prominent economists played a critical role in crafting environmental policy at the national level. Some domestic scholars felt a deep sense of helplessness and frustration partly because they had neither the obligation nor the right to discuss state-designed policies publicly. Yet a few well-regarded economists influenced policies devised for environmental protection. Moreover, their attention to sector-specific management issues brought early policy attention to industrial pollution. Their studies often shaped the formulation of official policies to address relevant issues. Their essays also publicized concerns regarding industrial waste emission and raised broader public awareness and stronger political will, leading to the final promulgation of the National Environmental Protection Law in 1979. Since then, environmental governance concerns have been integrated into various state-led plans for national economic-social development.

During the 1970s and 1980s, most efforts at containing pollution, as proposed by local economists and government employees in Hubei and nationwide, attempted to integrate environmental management within the economic policies of the PRC. Most of these experts raised environmental issues, though they were economists by training. One can further examine the scholarly contexts with methodological references for this multiple-disciplinary study and the earlier literature. A short survey of local economist scholarship reveals how the perspectives of state policy experts gradually shifted to take account of the need for environmental protection through the regulation of industrial waste emissions. It also traces the hidden logic leading to China's formulation of environmental policies with economic rationality. These economists played an influential role in environmental policy designs as the basis of subnational policy implementation. Some cognitive steps, along with environmental protection thinking, marked the emergence of environmental economics and other environmentalist thought elsewhere.

Ecological and environmental economics are discussed herein as interchangeable because both fields have largely overlapped policy design and implementation. Both fields once saw natural resources as free or inexhaustible. Neo-classical economic theory assigned no value to production inputs other than labour or costs. Departing from the neo-classical framework toward a more pluralistic evolutionary path of the profession of Chinese economics, some local economists argued that the non-renewable nature of many resources required new valuation methods that were not commensurable with money-measured capital.[2] Ecological economics also promotes socio-economic as well as environmental sustainability.[3] Ecological or environmental economics in the PRC was overshadowed by neo-classical economics, which made a triumphant comeback after the 1980s when public and academic attention began to focus on policy hinging on the "externalities" of environmental problems.[4] Nonetheless, core themes in the overlapping economic subfields generated debates among Chinese economists, including those who had first expressed concern regarding the ecological costs of economic development much earlier.

In the 1980s, Chinese economists who shared environmentalist concerns began to form small, low-profile academic communities. To promote research collaboration, they initiated cross-disciplinary platforms. In February 1984, they founded the Chinese Ecological Economics Society (CEES) in Beijing, under the joint sponsorship of the Chinese Academy of Social Sciences (CASS) and the State Forestry Administration of China. The secretarial office of the CEES was established earlier in CASS and was hosted by the Rural Development Institute. Thus, balancing rural or agricultural development with other economic sectors was the priority research interest of local economists along with other social scientists. Another national institute is the China Society of Territorial Economists (CSTE), established in Tianjin in 1981. Both the CEES and the CSTE are registered as non-profit organizations, and they recruit members from universities, institutions, and governments. Members are sought from both social science and natural science fields, ranging from ecology to philosophy. Provincial branches of the CEES and the CSTE formed loosely integrated regional societies, projecting the lack of an overarching theoretical framework for Chinese ecological and environmental economists, probably along with some political economists who were concerned about rampant pollution impairing economic growth.

According to a short review published by Fugu Cheng, a university professor and senior advisor to the Beijing government in 1983, the rise of ecological economics in China indicatively resulted from scholarly perceptions of critical environmental decline.[5] Western scholars focused on environmental economics as the global environmental movement was born during the 1960s and 1970s. This new subfield in China's academic circles, like environmental history, was not imported but expanded out of the traditional domains of economics, while being continually confronted by practical problems such as effective pollution regulation. Many Chinese experts tried to distinguish the field from other subdomains. For example, Cheng argued that Marxian principles for political economy shaped the dominant debates in China, as they also sought to resolve ecological deterioration. Cheng implied that ecological perspectives had already been discussed within Chinese economics since a few famous Chinese ecologists had already explored ecological economics. This trend had occurred during the 1920s and 1930s before the field was rediscovered after 1979 (Cheng declared that China's normative studies in ecological economics emerged in the late 1970s).[6] In short, this field featured inter-disciplinary research and focused on applied areas, but only or merely in a theoretical manner.

Tian Shi, a former professor at the Zhongnan University of Law and Economics, also applied a historical perspective to articulate a Chinese version of ecological economics by examining the origins, dilemmas, and prospects of the field. His research approaches and conclusions are more compatible than Cheng's with the disciplinary paradigms of Western economics. He remarked that most Chinese economists, particularly the younger scholars of the post-Mao generation, were willing to extend their disciplinary horizons. Chinese environmental historians noted that Western environmentalism in the 1960s developed as a social movement rather than from the regulatory role of governments. Based on Shi's accounts, also from the late 1970s, some reputable Chinese economists introduced ecological economics with new theoretical frameworks and research paradigms to China. Explaining why ecological economics had failed to influence policy effectively in 1970s China, Shi argued that the policy was overshadowed by economic conditions and political struggles, partly attributed to the ruling position of Marxist political economy after 1949.[7]

Shi's discussions drew primarily from published volumes rather than scholarly articles. Only a few scholars had the privilege of publishing in the leading Chinese economics journals within such a narrow scope. Among the promoters of environmental economics in its first phase (1980–84) introducing ecological economics, Shi singled out a significant figure for praise, Dixin Xu (1906–88). Xu's work in the early 1980s had also called for investigating how Chinese economists managed the discomfort of environmental damage and official neglect. Shi noted that open debates helped clarify the localized interpretation and policy-making processes of relevant fields. A flaw in his review of the literature would be his rigid periodization of the evolution of ecological economics in the PRC. His overview lists some meaningful events, particularly stressing the interdisciplinary nature of reality-based economic-environmental policy issues.

China's most respected economists published in *Jingji Yanjiu*, or *Economic Research Journal (ERJ)*. After its re-establishment in 1978, this top-tier periodical published a series of academic articles on China's environmental problems and economic agenda. Some of these articles are primary sources for examining China's public policy discourse on environmental protection policy. Many reputable economists contributed insights and follow-up responses to the *ERJ* discussions. Almost all Chinese scholars relied on government sponsorship, a fact that could compromise their academic integrity. Around 1980, Dixin Xu and Mingzhi Chen were among the few scholars who publicly addressed the reality of environmental protection problems. Chen recommended balancing ecological costs with economic and technical efficiency, while viable governance solutions should be environmental protection legislation with efficient policy execution. Xu advocated the integration of ecological economics into the state development agenda, officially labelled as the Four Modernizations in 1975, and encouraged prioritizing ecological equilibrium over economic equilibrium.[8]

According to Xu, ecological economics and territorial economics were closely linked. He explicitly stated that more emphasis should be directed to land management issues and that ecological economics should also be known as environmental economics to mitigate self-enforced restriction to some extent. Thus, when it came to terminology, Chinese economists did not show a preference for either "ecological" or "environmental" economics, which may have helped promote a more comprehensive foundation for academic collaboration beyond traditional domains of economics. Before introducing ecological economics in China, a few scholars, including those mentioned above, had called for more reflective examination of economic history, though initially they subtly stressed ideological or political aspects rather than ecological concerns.[9] These arguments framed China's economic growth from a deterministic perspective, from the revisited history of peasant rebellions, resource allocation, commercialization, and industrialization, and whether or not Chinese "sprouts of capitalism" had appeared in imperial times.

Considering the later policy practices of environmental protection, the studies published in the *ERJ* merit further interpretation from a historical perspective. Dixin Xu served as vice-president of the Chinese Academy of Social Sciences (CASS) and was a senior consultant to the central government. He was also the chief director of the Institute of Economics in the CASS, which has hosted the *ERJ* since 1955, except during the Cultural Revolution.[10] In addition to its discussions of environmental and economic policies, the *ERJ* printed a sequence of debates on China's economic and social reforms. Emphasizing theoretical applications with attention to policy implementation at the local level was reflected in economists' debates on rural reforms in the 1980s. For example, debates about how to redirect China's traditional agrarian economy by focusing first on reassessing its modernization path for new industrial policies resonated with their calls for economic reform.[11]

I consider it relevant to interpret the hidden logic of economic measures in Hubei's local environmental policy practices. As indicated in many records of environmental governance in Hubei, warning-first measures – or warning-only measures, as they might cynically be described – had been systematized yet with unsatisfactory results. The warning-first approach was used in many anti-pollution campaigns before the late 1970s. Once the HEP was established, administrative fines replaced the warning notices issued by industrial bureaus. Administrative red tape and inter-sectoral disputes continued to hamper regulation efficiency during the early 1990s.[12] At the national level, issuing a warning to a polluter was a vital component of the industrial policy. A series of national directives were abbreviated as regulatory codes on industrial emissions.[13] These directives instructed lower-level officials that all factory projects must be synchronized with the corresponding stipulations for design, construction, and operation in all related environmental protection projects. Failure to implement the pollution control policy would lead to symbolic economic sanctions. The codes of the industrial regulation policy were still based on the established principle of prioritizing industrial development.

Reports on industrial pollution in Hubei demonstrate that both national and subnational policy makers sought to avoid any risks from social or economic disruptions that would affect their political careers. Therefore, these individuals made prudent decisions in cases where egregious pollution meant that industrial production should be suspended or even terminated. Economic development was expected to gradually diminish the adverse effects of industrial waste by-products, but finding the optimal balance between economic growth and pollution-regulation goals was not a significant consideration for official decision makers. Economy-first justified the use of warning-first measures in environmental governance. Central and provincial planning committees should be ranked among the most powerful agencies in the PRC government system.[14] All the investment plans for pollution containment must first be reviewed and approved by the planning committees at various levels. At the same time, local environmental agents were required to produce environmental monitoring reports. Without these reports, no one could legitimately authorize investment budgets and allocate funding or assets.

The industrial waste control procedures included ten protocols that were to be followed to access investment funds to contain pollution. The first specified a complete procedure-filing system of task assignment and stage-control assessment; the second stressed site design and selection for any new industrial investment project; the third required monitoring of emission control, which was distinctly planned for existing or new projects. The fourth to sixth regulations related to the stages of design, construction, operation, audit, and approval, as supervised and operated by the appropriate government offices. The seventh stated that all new projects must firstly have demonstrated scientific experiments of effective emission regulation, except those that failed at the current technology level but were still assessed as "urgent" state projects. The eighth and ninth called for more active cross-sector collaboration by other government branches, including planning, construction, public finance, urban facility, and public health. The tenth and concluding protocol briefly noted that the preceding nine procedures would apply to all socio-economic units. A few specific terms created loopholes to get around the seemingly restrictive codes. In particular, the seventh regulation provided flexibility and de facto accountability-free judgment in policy implementation.[15]

Given the frequently shifting or vague standards regarding the implementation of policy, environmental and industrial officers would inevitably have been confused about how to proceed. As noted above, many archival records from Hubei show how provincial and county-level bureaucratic weakness hindered the regulation of industrial waste emissions. These failures demonstrate the relevance of administrative weakness. Most cases of industrial pollution followed a pattern that reflected the fragmentation of administrative responsibility among various offices and jurisdictions according to area and hierarchical rank. Soon after they assumed responsibility for environmental management, environmental agencies seemed to have no better policy choice than to apply economic rationality in practices that legalized, or priced, the right to emit industrial waste.[16] In addition to being morally objectionable, this strategy seems like downplaying the right thing and sanctioning the wrong thing.

Political economists who are interested in moral economy argue for the need for specific case studies, such as E.P. Thompson's study of eighteenth-century England. As he notes, "The confrontations of the market in a 'pre-industrial' society are, of course, more universal than any specific national experience. And the elementary moral precepts of the 'reasonable price' are equally universal."[17] Aware that people's underlying moral reasoning can be difficult to extract from their rhetoric, three Norwegian economists attempted to determine whether emission-trading schemes were seen as moral. In 2011, they designed an experiment in which lab participants (undergraduate students) played a game involving stickers. Any participant who used a sticker gained at the expense of the others, and ultimately with net harms for everyone involved. The game mirrored the long-term effects of tradable emission quotas, though the researchers did not reveal this or mention any climate/environmental context so as not to trigger pre-existing attitudes about the ethics of emission-trading schemes. An overwhelming majority of subjects did not hesitate to use their stickers, even though later questioning by the researchers revealed that they were strongly opposed to the real-life trading of emission quotas, which most of them saw as immoral.[18]

Moral discussions on the state policy strategy of granting pollution rights are related to another issue: human population pressure on nature. This subject seems especially relevant for China, where a strict population plan policy of one child per family was pursued from the late 1970s to October 2015.[19] Examining endogenous responses of a population to environmental policy, David de la Croix and Axel Gosseries address the economic-environmental tension via two practical ways to achieve low polluting emissions: reducing per capita production or population size and containing pollution either through Pigovian taxes (the taxes on any market activity whose costs are not reflected in the market prices) or through tradable quotas. In this economic model (or hypothesis), fertility, output, and pollution are all arguably assumed to be the endogenous factors, whereas taxing output to regulate pollution might delay a demographic transition, worsening the welfare of future generations.[20] To put a cap on reproduction or to recognize pollution rights might seem equally morally controversial, but both could be effective for environmental protection.

As a policy, the permission to pollute was an adaptive measure to deal with the disappointing results in regulating industrial waste emissions. Calls for granting such pollution rights were presumably all issued from governments to individual enterprises that produced emission waste. To explain this situation more specifically, it is essential to understand the structure of enterprise ownership in the political and economic institutions of the PRC. Industrial emission policy fell within a legal grey area, where vague boundaries existed between public and private domains. Most provincial environmental agents demonstrated a preference to avoid public discontent. Their anti-pollution regulation initiatives were characterized by economic reasoning. Given the existing administrative framework and political setting in China and the institutional constraints, they would require the ability to grant the right to pollute to avoid antagonizing established political and economic interests. This solution was seemingly more effective than earlier policy instruments, which were based on toothless, inadequate, or at best ineffective administrative warnings.

Soft budget constraints, as a concept, are used to delineate social issues deeply mingling with managerial responsibilities delegated to state-owned enterprises plagued by rent-seeking corruption and bureaucratic weakness in a planned economy.[21] Two well-known economists, Justin Yifu Lin and Guofu Tan, examine conflicts of interest in China's state-owned enterprises with policy burdens, including industrial emissions.[22] In their game-theory analysis, political conditions are settled in theory as an externality arbitrarily defined, including the factors (or effects) that are out of the scope of pure economic reasoning, arguably. The HEP reports showed reluctance to arouse political antagonism but sought to balance inter-sectoral economic development. The HEP had to resolve policy issues by economic measures rather than mobilizing political pressure. This appears to have been an interim solution. As more official and public discourse was gradually depoliticized later (and hopefully irreversibly), many Chinese state agencies incrementally replaced Maoist ideology with economic goals in Hubei and elsewhere.[23] If granting the right to pollute could put effective regulations in motion, this approach was at least better than empty warnings.

Economic Policy Predicaments with Quantitative Paradigms

In recent decades, qualitative and quantitative approaches have been introduced into environmental-economic policy studies with trade-off analytic paradigms. With the reform programs announced in 1978, Deng Xiaoping prioritized the position of science and technology in the national agenda of market-oriented economic reform. In 1978, some highly regarded Chinese economists began to support a leading role in applying science and technology to the creation of economic growth.[24]

Among them was Yushi Mao, later the 2012 winner of the Cato Institute's Milton Friedman Prize for Liberty for his influential work in advancing classical liberalism and free-market economics in China. In 1982, Mao analyzed three fundamental problems in environmental economics. His work promoted a few principal viewpoints embedded in China's state environmental protection policy. Mao stated that "reasonable" standards of environmental quality should be adapted to or should "correspond appropriately" with existing levels of economic development for all levels of government. He advised that an index system should be first developed to assess the macroeconomic impact on environmental protection, while proposing a quantitative model of calculating the total factor pollution coefficients of products by applying the Leontief input-output matrix model.[25] However, evaluating the long-term, evolving effects of the integrated implementation package of environmental policies would be a challenging task for empirical economists even today. Here are Mao's most frequently quoted words: "Adapting environmental protection standards to levels of economic development." This economic principle of "proper correspondence" has been invoked by governments to defend their failures to implement environmental policies.

Mao built a concise model of pollution coefficients, which can be used to justify implementing a "pollute first, clean up after" policy since all ecological costs could be measured in "accurate" economic terms. Financial compensation is awarded after certain economic development stages have been reached, thereby assuming a pragmatic perspective of technological availability.[26] Accordingly, the overall ecological costs of both production and consumption activities would be broken down and assigned to specific industries. The Chinese authorities already had sole state control of most industries, including the energy sector and most of the heavy manufacturing industries established during the 1950s to 1970s.[27] Most state-owned sectors would presumably endorse having such pollution indexes first designed by policy specialists and industrial engineers. However, Mao also admitted considerable practical difficulties in applying his mathematically sophisticated model, especially without any solid long-term monitoring data to support such analysis. Therefore, his model was useful as a theoretical or conceptual guide to quantifying pollution problems. But Mao's proposal to begin quantifying ecological costs gained widespread acceptance by economists and officials nationwide.[28]

The statement "placing the task of environmental protection into the framework of economic development" reflected a consensus in a group of elite Chinese economists and strongly influenced the subsequent trajectories of environmental policy in China. According to Yushi Mao's viewpoint of ensuring that an appropriate correspondence existed between development and environmental protection, economists must answer a fundamental question: What level of environmental protection is viable? This trade-off principle became ubiquitous when local environmental protection state-agencies attempted to resolve existing or potential tensions arising from prioritizing development over environmental protection. Mao's influence concerning China's environmental protection problems faded after that 1982 article in which Mao proposed quantifying pollution loss first in the *ERJ*, as his academic interests shifted focus. Recently, he has returned to concentrating on developing energy policies and continuing with his long-term work on economic morality and rural development policies. His ardent supporters, including many Chinese government state agents, have paid close attention to his wide-ranging work.[29]

Although they echoed the thought of some Western economists, these theory and policy debates among Chinese economists did not cite their Western counterparts during the 1970s and early 1980s. However, this situation gradually evolved to include more foreign expertise for some environmental policy debates. In recent years, Chinese scholars, arguing that neither the free market nor the visible hand of government should assume complete responsibility for environmental protection, have examined how to adjust economic measures to build a policy framework to balance efficiency and justice.[30] In the 1990s, Chinese economists referred to Western scholarship to model policy impacts, mostly following suit still at the macroeconomic level. These academic discussions among economists who were concerned with environmental policy implementation at local levels have increasingly favoured environmental (or political) fairness over economic efficiency. Since the 1990s, Chinese scholars have increasingly called for critical reappraisals of national or regional ecological protection policies to advance a more sustainable economy.[31]

Chinese economists were not influenced as much by the arguments of their Western colleagues before the late 1980s, apparently content to keep their expertise relevant in a limited scope despite similar environmental policy predicaments. Awkwardly for economic researchers who presumably played a policy-guiding role in most economic activities, China's relative global isolation has gradually evolved with its quickly expanding market-oriented reforms. A growing volume of environmental research conducted in China with policy recommendations also addressed emerging environmental justice issues. To balance environmental justice and economic efficiency, some scholars further examined the significance of China's environmental protection within international or domestic contexts.[32] Others focused on local research, including specific cities or rural areas. Some studies explored various aspects of land management and constraints on agricultural development in post-Mao China.[33] Chinese and Western economists have patently reached a consensus: All economic activity ultimately depends on a finite resource base. Institutions should thus provide sufficient incentives with policy design protecting ecological resilience and should treat this goal as a sign of progress in sustainability.

During the past few decades, economists in China and the West have created a rich body of literature with theoretical models while realizing that they may not perfectly apply to the real world. Incomplete information would prevent achieving a theoretically optimal state. This reality simply implies that environmental policy implementation must be adaptive. In 2012, Ali Douai, Andrew Mearman, and Ioana Negru published a discipline-wide overview of economics of the environment and sustainability.[34] Their review attempts to summarize the state of play in the economics of the environment from a still progressing conceptual map of contemporary ecological economics around the world. They examine contributions from traditional heterodox schools of thought. Their discussion reveals close connections between distinctive groups of scholars, particularly between ecological and heterodox economists. Early studies of ecological economics in China from the late 1970s have displayed the vigour and variety of heterodox approaches.

In reviewing the economic literature discussed above, I found no solid evidence to suggest that most environmental or ecological economists of the PRC had fully developed adequate guiding environmental protection or industrial waste emission policy designs within the existing economic theoretical frameworks. A myriad of high-ground economic principles appears either too theoretically abstract or too empirically complicated for applying more concrete policy measures for the purpose of pollution abatement. Chinese decision makers have positioned the national environmental protection policy as a critical part of the national economic policy framework since the later 1970s. A set of heated debates concerning the emergence of ecological and environmental economics in the PRC concentrated on what principles the state should apply to its environmental policy designs. During the transitional period from the 1970s to the 1980s, empirical studies by economists on the formulation of official environmental protection policies were filled with theoretical compromises. Articles published in the *ERJ* from the late 1970s to the early 1990s helped readers to understand the environmental policy weaknesses of state agencies and consoled them with more viable measures.

Between the late 1970s and the mid-1980s, most articles in the *ERJ* seemingly admitted that certain normative researches in China remained in the preliminary stages, but the governments did not have the luxury of waiting and so resorted to the trial-and-error methods for relevant policy experiments. This rising/re-introduced subfield of normative economics in the PRC has inherited some substantial influences from political economy. Economists have often articulated that environmental challenges would unavoidably accompany modernization campaigns. Confined to the Chinese political system, economists had neither the obligation nor the right to discuss official policy in the public space. Chinese scholars have been cautious in dealing with politically or ideologically sensitive topics. Some state authorities also appeared to be concerned that academic debates on environmentalism might encourage environmental politics. Importing a Western preoccupation with citizens' rights might undermine the legitimacy of the CCP leadership and the political autonomy of the PRC. The above scholarship review illustrates how some general principles or ideas help explain policy failures resulting from seemingly half-hearted pollution control policy measures.

A few elite economists sounded some early warnings about the PRC's industrial waste emission regulation and environmental protection policies, particularly the reappraisal of policy measures and developmental patterns. Their proposals for national environmental policies were filled with theoretical compromise. Their studies presented remarkably influential efforts to reconcile the traditional policy study approaches of normative economics with more quantitative methods. The lack of convincing economic metrics for assessing the environmental damage of pollution, along with the difficulty of identifying both the violators and the victims, also impeded effective industrial waste emission regulation and proper environmental protection policy design and implementation for the goals of protecting local ecosystems or furthering socio-economic justice. Nonetheless, the handful of reputable economists did speak out publicly. Their academic discussions demanded more recognition of the pressing ecological decline caused by manufacturing and poor pollution regulations.

Evidence of early failures in regulating industrial emissions in Hubei can manifest the essential role of adaptive methods with pragmatism to justify the right to pollute. In an internal report in 1973, Hubei proposed to the central government that it initiate more provincial propaganda to promote environmental awareness. Addressing the role of subnational leadership, the report aimed to motivate more consensus about environmental protection in Hubei.[35] Also in that year, environmental officers and directors of state-owned enterprises, a startling total of ten thousand people, were summoned to attend a grand provincial conference at Wuhan. Its top priority was to raise awareness of environmental protection in Hubei and to achieve policy consensus among officials of provincial and lower-level government agencies. Maintaining industrial waste emission regulation incentives was another long-term issue in environmental policy implementation. In the short term, local environmental protection state-agents of Hubei needed convincing measures to establish such governing mechanisms. This public responsibility for environmental conservation would require new or reinvestment projects and restrengthened initiatives. In 1998, Eduard Vermeer observed that environmental awareness was the key factor in achieving remedial policy solutions. As he concluded,

> The main challenges for China are to raise environmental awareness, train and expand a core apparatus of experts, promote participation in environ-

mentally-friendly policies and activities, and phase out wasteful enterprises using a combination of legal and economic measures ... The question is to what extent China is prepared and able to accept a temporary reduction of production activities for the benefit of qualitative improvements to its production structure and the quality of life for future.[36]

This insightful observation accords with what environmental protection state agencies documented in Hubei itself. For instance, in 1981, the Hubei Provincial Bureau of Metallurgy relayed a series of reports on the campaign to raise environmental awareness with a period of central agency directives. As a result, professional training workshops were sponsored by research institutes, including the Hubei Committee of Atmospheric Research and the Hubei Environmental Protection Committee.[37] Nonetheless, one can also refer to a comment from Vaclav Smil, written in 1984, that "the best outlook is for some gradual improvement and the prevention of further major degradation in key sectors and areas."[38]

From a shortlist of locally reported instances to a much wider monitoring area of general regulation, which still featured a point-control strategy, a notable shift of relevant policy implementation for industrial waste emission regulation gradually evolved from the early 1970s to the mid-1980s in Hubei. These findings have highlighted the government's administrative weakness, particularly in subnational bureaucratic arrangements. Compared to other restricting factors on economic activities, moral restraint did not play a positive role in the space or time examined. Adequate environmental protection would demand active policy implementations and people's commitments to coordinating with local state agencies. As mentioned above, the HEP struggled to acquire both necessary staff and state funding from its earliest days, gradually gaining status within the provincial bureaucracy.

A slow but progressive pattern of increasingly formalized regulation measures enforced by the HEP and its subordinate agencies across cities and villages also occurred. However, most of their measures remained inadequate, with similar weakness inherited from earlier decades. Taking the first step of issuing an administrative notice served as a warning to polluters. The notice was not closely connected to any meaningful penalties, given that polluters who breached regulations paid their fines and perhaps made a few minor improvements but essentially continued as before, with the result that the cycle simply repeated itself. However, the legitimation of administrative punishment was not officially approved until 1989. During that year, the HEP started to use legal action to implement policy.[39] As part of this transition, subnational governments like those in Hubei focused on economic growth. Efficacy issues would challenge low-ranking officers to achieve their long-term pollutant abatement targets. Financial punishments were criticized as too mild, so the effectiveness of regulations remained disappointing. This fragmented environmental policy implementation can be associated with the statist principle of promoting economic development, encapsulated as "develop first and clean up later." Environmental costs were placed on a de facto secondary agenda.

Pierre Desrochers examines the idea that most environmental regulations, once properly designed, can simultaneously improve both environmental goals and industrial competitiveness.[40] Stemming from the Porter hypothesis, the idea also advocates so-called win-win innovations to reduce environmental damage and increase industrial profits. Desrochers surveyed some little-known literature published in the late nineteenth and early twentieth centuries, which discusses the impact of market incentives on the development of valuable by-products that were recycled from industrial waste. Industrial experts and economists opined that clean production should be triggered by legal actions based on private property rights and government regulations. On the basis of some critical historical writings, Desrochers also suggested that the best way to craft well-designed environmental regulations would be to return to a private property rights approach to mitigating pollution problems.[41] Although this scheme might have limited usefulness in China's state-led system, many scholars would generally agree on including as many stakeholders as possible in industrial policies and relevant decision-making.

Many Western observers might feel justified in concluding that China's environmental bureaucracy has been weak simply because Beijing preferred to keep it so. Thus, the PRC state agencies often ignored environmental concerns simply because it did not make pollution abatement a priority for officials and party members who held government positions.[42] Such views, focusing most on the economic development agendas or results as typically noted since the 1990s rather than on earlier evidence at the provincial level, would be confirmed or rejected through exploration of how institutions were established and adapted to national policy. Many local economists and policy scholars expressed a stronger interest in environmental protection policy incentives than their Western counterparts did.

The lack of commonly shared incentives, as in proactive actions for environmental protection, was also recognized early on by officials in Hubei. In April 1970, an urgent report by the Hubei Revolutionary Committee on industrial effluent pollution stressed that all state agencies should take immediate action to halt the poisoning of Hubei's river and lake systems.[43] From the early 1970s, taking the political route of mass mobilization, as one of the most familiar strategies available to bureaucrats in the Chinese communist system, had already been considered by the provincial government and CCP leaders in internal talks for improving the provincial environmental protection agenda.[44] These archived government documents presented concerns about damaging the political alliance between the rural population and urban industrial workers. Inter-sectoral conflicts received internal attention from both senior and lower-level officials. Nonetheless, one should probably not view the subnational state agents of environmental protection as receptive to the political implications of their daily work or as advocates of a political transition for China before more remarkable economic development had been achieved.

Interpreting the Right to Pollute: Why Ask for Permission?

In 1974–75, the Provincial Leadership Office of Environmental Protection, a predecessor of the HEP, launched its network of environmental protection monitoring stations. Its early tasks were supervised by the Hubei Construction Committee. At first, the Hubei Provincial Department of Health was strong-armed into playing the lead agency role in these tasks. Still, the health agency divested itself of taking responsibility for environmental protection. Later, central directives confirmed the formal establishment of the HEP, initially intended to serve as a branch of the public health system in 1975.[45] The HEP's network of stations was affiliated with the existing public health network around 1979. Its regulatory proposals were reported to the provincial state planning agency.[46] When the HEP began to operate, its staff and funding were first assigned at the local level through administrative transfers from other government offices. The EPMS network stations' budgets at the municipal and county levels followed the Industrial Infrastructure and Public Transformation Facilities rules.[47] Beijing's final approval of the HEP was received in 1977. Central authorities referred to ecological deterioration, most closely associated with industrial waste emissions, as an obstacle to continued economic development.[48]

Anxiety and anger in Hubei are reflected in a letter from Daye county governmental officials to their superintendents in 1980, responding to the news that a staggering fine of 2 million yuan/RMB for violating emission regulations had been imposed on the Daye Company.[49] As the writers noted,

> The situation in Daye county, affected by industrial pollution, has become dire. It has directly affected the development of agricultural and machinery-manufacturing production, the people's health, morale levels in achieving the Four Modernizations among the people and officials, and the alliance between farmers and industrial workers ... While public directives are continually "paraded" in government offices, are they [government leaders and agencies] not concerned that the people imperilled by "public hazards" might rise and "parade" in rebellion?[50]

This comment evoked a challenging question: Can such a provisional policy solution resort to the practice of granting permission to pollute (which had already emerged from the grim realities) during the search for satisfactory policy solutions to environmental problems? The lack of adequate policy measures to contain or manage industrial pollution, rather than merely monitoring emissions, was formally recognized by both central and provincial agencies. In October 1978, three branches of the Beijing government, the State Planning Commission, the National Economic Committee, and the National Leadership Office of Environmental Protection (predecessor of the Ministry of Environmental Protection), co-issued a national directive listing industrial enterprises that had been ordered to reduce their pollution within one to three years. In doing so, these agencies created the first list of 167 industrial projects with the label of *xiangai*.[51] This term, abbreviated from *xianqi zhenggai,* refers to a formal warning notice that demands "prompt rectification within a definite time," a punishment that was regularized during the following years.

This xiangai list was soon allowed to be circulated among offices at the provincial level, and it consisted of quite a few senior-ranking enterprises in the state-owned industrial groups, including the Wugang Group and the Wuhan Number 2 Chemical Factory, both in Hubei. According to the directive, the list would serve as a preventive measure. It also reflected an underlying logic of prioritizing economic development. Being listed might seem to be a black eye for polluters, but the reality turned out to be otherwise. Far from imposing any meaningful economic penalty on violators, the xiangai system ultimately rewarded them. Other lists were drawn up, and, ironically, state-owned enterprises competed to be included on them. The reason was simple: being named on such a list made a facility eligible for an investment reward. As per central directives relayed to factory managers, polluters were eligible to receive investment funding for technological upgrades. Being allocated an investment quota might help them qualify for further resources to support an expansion of production. For the manager of any state-owned enterprise, undertaking a few basic calculations of potential costs and returns would reveal the obvious fact that getting listed as a major polluter made good economic sense.

Two messages were delivered to the managers and their official supervisors: first, investment in emission abatement technology was an acceptable solution to industrial waste problems, and second, failing to take remedial action and continuing to pollute might result in financial penalties. The xiangai policy contained a threat that operations could be suspended given the problems were resolved at least tentatively. As noted above, there were numerous cases in Hubei where facilities reverted to emitting waste as soon as they reopened.[52] But given the value of time, even for industrial managers in a planned economy, the policy contained an incentive to make enduring efforts to regulate emissions by providing pollutant-abatement device investment funding.

When examining public policies for environmental protection, Chinese and Western economists have applied similar analytical paradigms, using the terminology of incentives. One can suspect that Chinese scholars and officers would pay much attention to learning foreign expertise for guidance in domestic issues. However, some general economic principles sounded much more appealing to Chinese policy makers. Readers may refer to institutional economics, a theoretical framework that is consistent with and complementary to standard neo-classical economic theories.[53] Because of discrepancies between private interest and public well-being (or damaging costs) mainly in both economic and social terms, the rent-seeking hypothesis would predict increasing transaction costs due to corruption among state agents. Conventional policy solutions from environmental or ecological economics have not differentiated between a comprehensive policy package or a single mechanism or golden rule concerning how public goods or services should be maintained.

Debates about pollution rights in Western academia can also be traced to the late 1970s and the early 1980s. Economist Randolph Lyon began to analyze systems of transferable pollution rights in the early 1980s. Policies recognizing pollution rights, including rights-auction systems, free initial distributions of rights, and alternative procedures for allocating rights, have increasingly attracted Western researchers and policy makers. In his simulation model with excellent mathematical skills, Lyon argues that a fundamental trade-off existed between reducing the financial burden of industrial waste emission rights purchases and eliminating incentives for strategic behaviour. Opportunities for the latter may exist following non-incentive-compatible procedures because of small quantities of pollutant emissions. Lyon concludes that, despite incentive compatibility issues, the rights allocation approach via a mode of free initial quota distribution by authorities, followed by free-market exchange, may present the most promising package of features for pollution management in terms of efficiency, equity, ease of administration, and certainty of outcome.[54]

Charles Howe and Dwight Lee discussed pollution rights in a 1983 article titled "Priority Pollution Rights," proposing that environmental variability in an economically efficient manner should feature the right to pollute in preliminary policy experiments of relevance.[55] Howe and Lee call this design a "priority pollution rights system," an adaptation of the "appropriations doctrine." The system seems to create greater flexibility with more efficient use of privately held information on technological and cost conditions than alternative instruments in adapting pollutant abatement efforts to the assimilative capabilities of the environment. Howe and Lee also opine that it includes the possibilities of undesirable forms of strategic behaviour in the market; moreover, such possibilities could exist under all instruments. The authors conclude that successful experience with priority water rights in the western United States boded well for similar systems of pollution control.

Similarly, Donald Ryan agrees that market-based incentive programs should help manage property recourses efficiently. Economists created a large body of literature featuring detailed theoretical models. However, real-world imperfections, such as incomplete information and transaction costs, make it unlikely that theoretical optimal states could be realized. This reality implies that environmental policy must be adaptive. Ryan notes that laws and standards, including all the relevant state-issued dictates and industrial codes, ought to be periodically reassessed, insinuating that no "terminal" decisions exist. Ryan thus suggests that theoretical treatments of waste emission management should experiment in the real world. Despite the theoretical and information deficiencies, he writes that a "marketable emissions permit program would be a worthy candidate."[56] Although Chinese environmental agents did not use the term "pollution rights," in Hubei they had already taken their first step toward permitting de facto *negotiable* pollution rights within a market of pollution "suppliers" and buyers of emissions permits.

My analysis takes a sympathetic tone regarding Hubei's environmental agents as a collective struggling to balance short-term and longer-term goals. It seems essential to consider their social setting when assessing how they addressed their policy goals. How they arrived at their bureaucratic jobs and their insiders' language and mindset are critical components that the interpretations of outsiders might overlook. As Peter Perdue points out, in the environmental history of the PRC, continuities and disruptions are two sides of the story; to neglect either side causes an analysis to lose its coherence.[57] Perdue's observation helps grasp the continuity of environmental policies to some extent. According to the principles of institutional economics, governments may take responsibility for supplying public goods and services (environmental governance in this case) and may do so at lower costs than would be incurred by private voluntary organizations. In the case of China, the leading role played by the state is deeply embedded in the strategies by which agents and citizens would all seek to adopt policies to meet long-term challenges in subnational situations.

While recognizing the distinctive features of China's political and social systems, we can nonetheless analyze political influence as a distorting effect. Among relatively recent researchers, Yu-Bong Lai has examined this policy issue by legitimately and fairly designing a system of tradable emission permits, including the justifiable distribution of initial pollution rights. Setting the initial rights or determining the number of tradable permits depends on the influence of interest groups. In the presence of political pressure, Lai shows that the distribution of the initial rights could significantly affect social welfare. In his model, revenues from the auctioning of permits are distributed to environmentalists and consumers; if the proceeds are refunded according to relative output levels, strategic effects may arise, and industries will attempt to shift rents away from other industries through refunds. In other words, various interests and perspectives of individual industries may create competing strategies for tradable emission permits.[58] Thematically related to some historical observations in the present study, Lai's theoretical framework calls for further debate. One may complain about the vagueness of such theoretical approaches resembling the policy solutions that Chinese economists proposed from the 1970s to the 1980s, as reviewed above. After all, putting a price tag on blue skies with fresh air, fertile soil, and clean water, which many of us take for granted as free and available to all, sounds so ludicrous.

Unlike the civil service, the media have played an increasing role in exposing pollution cases in the PRC. In recent years, it has become clear that public officials often take steps to silence and suppress public outcry whenever yet another case of large-scale ecological damage surfaces. Two relatively recent incidents are illustrative here. On November 1, 2006, the Chinese media reported an accidental ammonia leak at the County Ammonia Plant at Dawu, Hubei, which claimed one life, left six severely injured, and thus required the emergency evacuation of twenty thousand residents. According to the media, ten tons of ammonia residue were accidently flushed into the Fu and Huan Rivers, creating a "potential" environmental hazard.[59] Hubei's regional news media concentrated on the efficiency of the emergency measures deployed to deal with the problem rather than on the seriousness of the spill. In their defence, most media are state-sponsored and would thus experience the same frustration and powerlessness as environmental officers, despite the expectation that they are bound to advocate environmental governance with political impartiality, social justice, and transparency.

A more recent case of river pollution aroused more political concern. On September 5, 2013, American broadcaster CNN reported that Hubei authorities had scooped up tons of dead fish that had been poisoned by ammonia discharged from a chemical plant on the Fu River. CNN noted that dead fish covered a forty-kilometre stretch of the river. Provincial authorities charged a state-owned company, the Hubei Shuanghuan Science and Technology Corporation, as responsible for the disaster.[60] Alarming incidents, like this "fresh" case, would naturally create public suspicion regarding the safety of drinking water. Concern for failing policy enforcement of environmental protection undermined the frequently announced "ecological modernization" goals under the current PRC leader, Xi Jinping. A lack of Chinese reporting about the September 2013 incident reflects Beijing's policy of suppressing news coverage that may damage public trust in the achievement of national environmental goals backed by local governance. Despite many updated policy goals, ecological modernization is still a vague concept waiting to be defined, given that its policy advances seem to lag far behind social and economic progress.[61]

As shown above, the problems of multiple pollution sources and causes help explain the predicament of provincial and sub-provincial environmental agencies. The Chinese state also played a critical role here. Therefore, the foundations for the regulation of industrial waste emissions were laid through a pragmatic but problematic approach of develop first and clean up later. As observed, Hubei's state agencies diminished their knowledge gaps and learned from experience. In the longer term, however, environmental governance gradually improved. Although industrial pollution is the main focus of this book, other cases in Hubei reveal that urban and rural communities situated close to industrial zones were also affected by increasing levels of household-waste emissions, typically causing eutrophication. Therefore, economic reasoning inevitably emerged around how to handle industrial waste pollution–caused costs rationally and how to provide damage control by compensating the victims of pollution.

Conclusion

Early policy experiments in Hubei's industrial waste emission measures reflect the complexities of many interlocking factors in China's economic development. Because food security was of such vital importance, a few senior officials in Hubei were assigned responsibility for mediating the frequent conflicts of interest between industry and agriculture. Most early environmental policies in Hubei primarily constructed a pragmatic management system that was based on vaguely outlined economic principles. In response to accumulating socio-economic pressures, a series of measures evolved into a process that essentially legalized the right to pollute. This reasoning, to adapt to state-endorsed development goals that prioritized economic growth, made sense to state-owned firms whose decisions about waste emissions were entrenched in the same principles. Furthermore, environmental protection officials were not provided with sufficient incentives to pursue effective solutions through mediation or other means. Given that their tasks were prioritized in correspondence with the economic growth rates in their jurisdictions, they could do little but pay off rural and urban victims of industrial pollution.

6
Updating Environmental Governance in Wuhan, 1986–90: Further Regulations for Blue Skies

In retrospect, most industrial regulation measures of the PRC consisted largely of either vague or overly generalized principles, though typically encompassed within an overarching governance framework of economic reasoning. Given this, when we read the reports written by the environmental agencies of Hubei or Wuhan, we need to examine them on their own terms rather than applying an approach in which their authors are seen as "Others." Their reports commonly recorded pecuniary compensation, trade-offs for economic growth, regulation efficiency, and inter-sector coordinated growth, to mention just a few topics, but all stressed calculations of economic benefit-loss and proper reparation to specific sectors or residents who were identified as victims of illegal emission incidents. Table 2 provides a timeline of events that help readers unravel the complexities and struggles of environmental governance in Hubei.

Regarding the evolution of local environmental governance in Wuhan during the 1980s, I consider it relevant to elaborate a few conservatively opportunistic opinions with its specific contexts and cultural settings. I examine its city history first, with its urban legacies featuring continuities and ruptures. Early city planning in Wuhan was much like that of other major cities in the PRC; thus, to some extent most cities largely resembled each other. The Wuhan city plan was updated at the beginning of China's

TABLE 2 Events correlated to Hubei's environmental policy evolution from 1952

Years	Event	Area of impact
1952–56	Jingjian Flood Diversion Project	Wuhan and Hubei
1957–58	Preliminary experiments with biogas technology	Wuhan
1958–62	Great Leap Forward Movement	Nationally
1959–61	Three years of famine	Nationally
1964	Reports of fishery grievances due to industrial effluent	Provincially
1966	More reports of chemical factories polluting Ya'er Lake	Wuhan, Ezhou
1966	Cultural Revolution begins	Nationally
1970	Hubei emergency directive for industrial waste control	Provincially
1970	Provincial Three-Waste Governance Office pre-established	Wuhan, Urban Hubei
1972	Water pollution report by the Hubei Provincial Bureau of Water Utility and Irrigation	Provincially
1973	Dong Lake pollution investigated	Wuhan
1973	County and municipal Three-Waste Governance Offices gradually established	Provincially
1974	National Leadership Office of Environmental Protection in the State Planning Commission established	Nationally
1974–75	Environmental protection monitoring stations further expanded into a few main industrial cities	Wuhan and elsewhere
1975	National directives of the Three-Waste Governance Movement	Nationally
1975	Provincial Leadership Office of Environmental Protection	Wuhan
1975	National call for the Biogas Promotion Movement	Nationally
1975	Hubei Biogas Office established	Wuhan
1976	Official end of Cultural Revolution	National
1977	Hubei Research Institute of Environmental Protection, the formal establishment of the Hubei EPMS system	Wuhan and elsewhere
1978–79	Dong Lake and Ya'er Lake cases reopened	Wuhan, Ezhou
1979	China's provisional National Environmental Protection Law passed	Nationally
1979	Hubei Provincial Bureau of Environmental Protection established	Wuhan
1979–80	Fuhe River pollution investigation	Wuhan, Xiaogan
1979	Daye county soil contamination investigation	Daye
1980	Municipal/county environmental protection bureaus	Provincially
1980	Wuhan Air Pollution Regulation	Wuhan Districts
1980	Dawu county soil contamination investigation	Dawu

Figure 10 Aerial views of Wuhan, 1985–89. *Top:* The three original towns of Wuhan – Wuchang, Hankou, and Hanyang – counter-clockwise from the top. *Left:* The Wuhan Marshalling Yard in the Jian'an District of Hankou. Intended for freight trains, it was built in 1955 mainly to facilitate the Wuhan Changjiang (Yangzi or Yangtze) Bridge. Still one of the largest in China, it was renamed as Wuhan North Marshalling Yard after several new yards were built in the city. | *Wuhan Municipal Gazetteer* (Wuhan: Wuhan Municipal Chronicle Office, Wuhan Publishing House, 1985, 1989)

economic take-off from 1979 that would be mostly accredited to its industrial achievements in the decades afterward.

As explored in Chapters 2 and 3, the city of Hanyang represented a notable case of China's earliest experiments with industrialization. To some extent, the Hanyeping Company was doomed from the start, as it would be both debased by insiders and raided by outsiders. Yangwu reformers such as Zhidong Zhang and Xuanhuai Sheng were the main advocates for industrializing Hanyang and are both irreplaceable figures in the history of the company. Although one cannot view their dominance as positive, they made great efforts to foster innovation. Some early policy efforts had entailed gradual progress in the institutional framework for industrialization. Despite their pioneering efforts to stimulate machinery production, the resulting over-dependence on a small group of elite decision makers would put their reforms at risk and increase the chances of fraud. Although critical in many ways, the role of entrepreneurship could not replace an institutional mechanism to promote accountability. More specifically, regarding corruption issues, a high level of trust within a small group of elites often suggests a low level of trust in society at large. In short, the overweening dominance of an elite minority of policy makers could also explain the governance failures in China's earliest industrial experiments.

The quieter districts of Wuchang are home to many intellectual communities. As a vehicle for preserving traditional and modern knowledge, Wuchang has gained a less secular-materialistic but more spiritual-literary reputation than its sister towns of Hankou and Hanyang, and it has a somewhat more privileged position than they. As a unified whole, they rely on mutually supporting urban functions. When outsiders settled at Wuhan, they brought economic and social credit with them. By contrast, many people, identified as local celebrities in the *Wuhan Municipal Chronicle*, moved to other cities with their personal achievements. "Self-strengthening" stories from industrial Hanyang, commercial Hankou, and cultural Wuchang have celebrated Wuhan's rich urban migrant legacies and cultural complexities with both bitterness and pride.

One can relate the resistant spirit of Wuhan to a well-known proverb, roughly translated as "poor mountains and evil waters make troublesome people." This once popular idiom is prone to blaming the residents of ecologically declining areas for their own governance problems. Readers may interpret it as taking the side of state agents rather than that of locals who tenaciously appeal against harsh conditions. Such rhetoric is replete with human-nature tension, as noted in many local gazetteers, though most would quote only the portion that refers to the mountains and the waters, and would diplomatically omit the comment about people. A derogatory label often refers to the people of Hubei as a "nine-headed bird," a monster creature from mythology, in connection with their unyielding attitude. When local state agents of Hubei interacted with civilian societies, certain grievances would be implicitly depicted. Asserting that adverse natural conditions will automatically propagate spiteful people is problematic, and such claims probably have more to do with failures in governance. Many life stories of Wuhan residents are grounded in cultural legacies and tenacity, valued by a diversity of local communities. This subculture would encourage engaging multiple stakeholders, and it would certainly help create better governance to manage pollution issues.

Regularizing Environmental Protection Work in 1986

In 1986, Beijing launched its seventh five-year plan, which ended in 1990. Also in 1986, reports on environmental protection became a standard feature of the socio-economic sections in the *Wuhan Municipal Chronicle*, which makes the period 1986–90 especially fruitful for study. Although other sections of the *Chronicle* had occasionally included discussions of local environmental matters, 1986 marks the launch of a regular subsection that systematically profiled them. The *Chronicle* office archived many official or semi-official reports written by government offices, though it appears to have printed what it saw as the most significant. Initially, the reports on environmental protection were part of a section titled "Urban-Rural Construction," whose subject was infrastructure. They achieved greater prominence in 1987, however, when the section title was changed to "Urban-Rural Construction and Environmental Protection." In the 1988 *Chronicle*, urban-rural construction regained its separate status under its original title, and environmental protection was covered in a brand-new stand-alone section and was officially known as "Environmental Protection." In the ensuing years, it appeared immediately below the section on urban-rural construction and the sections on other economic or industry sectors.

In 1987, the *Wuhan Municipal Chronicle* printed a report on the Donghu (East Lake) Sewage Interception and Treatment Project.[1] It noted that by the end of 1986, part of the project was nearing completion. Moreover, the project was part of the Donghu Water Source Protection Initiative, which had begun in August 1984. It was the first urban sewage treatment construction project in Wuhan since 1949. Its first phase (assessed in 1986) would intercept, collect, transfer, and treat local industrial effluent and household sewage from three districts in Wuchang. It would cover an area of 12.9 square kilometres and would affect about 263,000 residents, as well as more than seventy factories and a long list of public service units, including schools, hospitals, universities, and research academies. Every day, the treatment facility would handle two types of effluent: 1,300 tons of industrial wastewater, a rather surprisingly low figure, and 47,000 tons of household sewage. Celebrated in the *Chronicle,* the project had a new sewage treatment facility designed to process 50,000 tons of effluent every day, almost fully finished, a new pumping station, and a twenty-four-kilometre sewer mainline. The latter was still under construction, and only four kilometres of the mainline had been completed.

Another environmental modification project of the mid-1980s was at the Huangxiao River in Hankou.[2] This watercourse, about 12.4 kilometres long, is wholly human-made; containing more than 75 percent of the catchment area of drainage water in Hankou, it was designed as a drainage conduit for lower silted land in the nineteenth century. It was at that time that the digging of this river helped drain the lower wetlands of Hankou, thus enabling the creation of its famous commercial centre. Before being supplanted by the railroad, the river was also a short-distance transportation route connecting Huangpi county and Xiaogang county, both of which border Hankou. As a result, its name gradually evolved by combining the first characters of each county name. Most early residents who lived along the river originally came from these two counties. Both are well known for their migrant bricklayers and artisans in cities across China today. The formation of the urban settlement along the Huangxiao was attributed to its cheap cost-of-living. After decades of continual silting, the neighbourhood became a large slum frequently afflicted by waterlogging. Encircled by new urban districts with a growing population, the Huanxiao was the dumping point for garbage and effluent.[3]

In 1982–83, two flooding incidents along the Huangxiao resulted in enormous losses – the direct economic cost caused by waterlogging alone was 500 million yuan. Local district governments proposed re-directing the river frequently silted through a box culvert in order to resolve the effluent discharge problem. A massive civil engineering project ensued, divided into two phases: The first one rebuilt the neighbourhood, expanding around the notorious dumping area, including the repair of factory-residential zones and surrounding district development. Starting in October 1985 and ending in June 1986, this phase laid 2.5 kilometres of box culvert and installed four pumping stations, connecting conduits, and various auxiliary buildings. It also completed the first 1.7 kilometres of a new road, Jianshe (Construction) Avenue, which were built on top of the covered river.

Wuhan's official narratives portray the case of the Huangxiao River as a local version of Longxugou, the famous "Dragon Beard Ditch," a Beijing slum that was greatly improved by the government during the 1950s. The story of the poverty-stricken district's rebirth was a model of socialist reconstruction for Beijing's urban poor. A well-known stage play, titled *Longxugou* and written by the prestigious dramatist Lao She of Beijing, is still celebrated today as a socialistic achievement, one that keeps alive the memory of Beijing in the early 1950s.[4] The initial phases of the sewage treatment and Huangxiao River projects were both finished in 1986, with their second-phase investment expected to begin in the next few years. According to the *Wuhan Municipal Chronicle*, both were the "first of their kind."

The Huangxiao river-to-road project was Wuhan's first use of a large-scale box culvert to manage effluent and sewage. In the mid-1980s, the city records attributed the problem of effluent pollution to insufficient investment in treatment facilities, a problem that largely overlapped with an exploding demand for mass consumption and production. In their completed forms, the sewage treatment and Huangxiao projects can be compared with the Donghu Water Source Protection Initiative, which supplied almost all of Wuchang with clean drinking water from the Changjiang River. In the decades after the 1980s, increasing numbers of roads were repeatedly torn up for the purpose of extending or replacing sewage pipes, unavoidably interrupting traffic and inconveniencing daily life, sometimes to the annoyance of residents, whose frequent letters of complaint were published in the local press.

Figure 11 *Top:* The mouth of the Huangxiao River before it was covered. It does not empty into the Yangtze River directly, but connected to a pump station in the Fu River that enters the latter in North Hankou. *Bottom:* Built above the Huangxiao River, Jianshe Avenue is one of Wuhan's busiest streets. | *Wuhan Municipal Gazetteer* (Wuhan: Wuhan Municipal Chronicle Office, Wuhan Publishing House, 1985, 1988).

On Jianshe Avenue, a line of skyscrapers was erected with street gardens, and the thoroughfare became the new financial centre of Wuhan, a Wall Street district that hosts a myriad of city, provincial, national, and international financial institutions. Rebranded as a central business district, it is also home to the British, American, Canadian, South Korean, and French consulates, with an embassy for the Russian Federation tentatively scheduled to begin construction by 2023. Wuhan today is probably much closer to its goal of becoming an industrial and commercial centre comparable to Chicago, yet with cultural and intellectual influence on the national scale.

A long-term deficiency in urban management and civil planning expertise had also contributed to some environmental problems that arose in the booming metropolis of Wuhan during the 1980s. A 1986 overview report for Wuhan's environmental protection work listed twelve main civil engineering tasks that had been completed during the year to contain effluent pollution. They ranged from water-facility and urban sewage grid construction to surveying and profiling industrial emission data. Eighteen facilities were awarded the title of "clean and pollution-free factory," because all had complied with environmental protection codes. In terms of institutionalizing Wuhan's environmental evaluations for new civil construction or industrial projects, the records are divided into two categories: newly strengthened codes for urban producers, and agricultural subsector structure policy adjustment recommendations for rural areas. In particular, the report marked continual promotion in sustaining rural biomass energy. Preparation work for waste discharge fee collection instead of fines presented a more promising perspective, in terms of regulating efficacy, than simply punishing the convicted violators of emission codes.

Another relatively short but attention-worthy entry notes,

> The account registry work for waste-discharge fees has reached a total of 1,238 local business or economic organization units on the city scale, presenting an annual increase of 492 newly registered and a total annual collection of 27.32 million RMB/Yuan; ... The city added 191 new pollution-control devices for 176 industrial pollution sources registered in 169 sites.[5]

Referring to these ten priority items of the twelve main civil engineering tasks, which were accomplished in 1986, the overview report listed two more items with a line of policy proposal reports (brief abstracts), including Wuhan's environmental prediction report for the year 2000, its seventh five-year strategy plan and policy design reports, and the urban noise-violation fine collection code with a city announcement for protecting local frog populations. The content of these government reports seemingly delivered cynicism mixed with cautious optimism.

However, the conclusion of the 1986 overview report admitted unsatisfactory outcomes in many regulation agendas for containing water, dust, and noise pollution. For instance, citing one example of initial goals not being fully achieved below national requirements, the 1986 report admitted that the water quality of some urban lakes in Wuhan continued to decline despite diminished acid rain and lower annual dust-fall levels. It noted that pollution was worsening in the Wuhan section of the Fu River and in Moshui Lake in Hanyang (*moshui* literally means "ink"). In contrast, noise levels in the crowded commercial districts of Wuhan, mostly in Hankou, had stabilized at a range of seventy-five to seventy-eight decibels, far above the national standard. These brief reports logged disappointing results in the mid-1980s, while more policy experiments underscored a pragmatic pattern to mobilize state agencies, residents, and civil societies.

These policy experiments advocated environmental quality as openly addressing a special form of public good or service. Like their domestic and overseas counterparts, Wuhan's municipal agents were designated the civil task of maintaining such public services and goods, all related to the work of city planning and infrastructure investment. These records in the *Wuhan Municipal Chronicle* covered dam-dike construction, water facility maintenance, city transportation, gas-fuel supply, public green space, street cleaning service, and residential development. The last item, residential development, gradually vanished from the *Chronicle*'s section on urban-rural construction after the marketization and privatization of the real estate sector.

From 1986, slowly but steadfastly, an emerging environmental protection industry appeared to gain independent status in Wuhan, in parallel with other expanding industries during this transitional era. A milestone event in the development of the industry was the establishment of the Wuhan branch of the China Association of Environmental Protection Industry, registered as a Non-Government Organization (a "civil society organization" as translated from the Chinese), on June 25, 1986. More than fifty enterprises in Wuhan were involved in its product sales or specialized

Figure 12 Noise pollution detectors on the street. | *Wuhan Municipal Gazetteer* (Wuhan: Wuhan Municipal Chronicle Office, Wuhan Publishing House, 1988, 1990).

products for pollution-reduction operations. The Hubei environmental industry employed twenty-six thousand workers, about 10 percent of whom were registered as engineers. It produced 276 pollution-control items. Monitoring devices and pollutant reduction equipment helped contain solid waste, wastewater, and noise. Competing for industrial emission control projects within and beyond the provincial borders, some local environmental businesses of Wuhan had won national awards since 1978, gaining national reputations particularly for water treatment plus pollutant detection. One example was the Wuhan Instrument and Meter Factory, established in 1965, which patented its technologies for water treatment. Some of its contamination-containing products were sold to Hong Kong and overseas, including to Pakistan, Tanzania, and Romania.

Wuhan's environmental protection agencies conducted a city-wide survey of industrial waste emissions based on a 1986 municipal investigation report. They paid the most attention to dust and effluent discharge. Municipal planners set 1985 as a baseline year for statistical comparison with later years. Eighty-one factories were identified as smoke-dust polluters, and the most affected boroughs, ranked by air quality, included Qingshan, Qiaokou (in Hankou), Wuchang, and Hanyang. Twenty-six factories were identified as major emitters of effluent, most of them related to the Wugang Group at Qingshan. In 1985, the factories under investigation accounted for about 92.0 percent of the annual GDP of Wuhan and about 97.0 percent of its total industrial energy consumption, mostly in the form of coal. Even when residential energy consumption was factored in, they still accounted for 91.78 percent of total energy consumption in Wuhan. The investigation detailed that only 38.7 percent of the industrial smoke-dust emissions and 40.23 percent of the effluent emissions had been adequately processed. The recycling rate for solid waste, at 83.84 percent, was seemingly the best managed, but a mountain of tailings and slag still remained from earlier decades. Whether precise or a rough estimate, these percentages provided reference points for future policy designs and regulations.

Another local event of significance was the publication of a five-year research project on pollution in Dong Lake.[6] In January 1986, the Hubei Provincial Committee of Science approved the findings of the project. As discussed in Chapter 2, the research report included some shocking findings based on continual sample monitoring of the polluted waterbody. The Wuhan Tongji Medical University led the investigation and confirmed the conclusions of other institutions. Their reports highlighted the health risks of long-term exposure to waterborne organic pollution caused mainly by eutrophication. Based on the reported entry items, three outcomes seemed relevant: first, the Hubei provincial government eventually decided to rely on the Changjiang River rather than Dong Lake as the source of drinking water for the surrounding districts; second, the city government started designing more specific codes for water facilities and kept conducting more of such policy experiments; third, a new research facility, the Wuhan Institute of Environmental Medicine, was jointly founded and administered by the National Bureau of Environmental Protection and the Wuhan Tongji Medical University in November 1986. A priority of the new institute was to help write a research report titled "Wuhan 2020 Environmental Prediction and Policy Strategy Research."[7]

The "Wuhan 2020" report created a set of environmental protection policy proposals that seem obvious now, including installing small-scale decentralized electricity generating stations and distributed natural-gas stations, curbing urban sprawl, replacing heavily polluting "sunset" industries, regularizing environmental assessment reports, and building more satellite cities. All these innovative ideas seemed compatible with the increasing engagement of civil societies and better public governance for coping with environment-human conflicts. Two follow-up reports cited the Wuhan Iron and Steel Company and the Wuhan Cement Factory to illustrate the recyclable economy, presenting them as good examples of attaining economic benefits via managing waste emissions.[8] The factory was an auxiliary facility for the Wugang and Qingshan township and local districts development. After two years of effort, it was awarded 30,000 yuan to reduce air pollution, which sounded like a public relations campaign and a similar award for Wugang, who earned annually 2.5 million yuan from recycling its dust emissions; the Wugang group gained 550,000 yuan. The gestures endorsing environmental protection sometimes mattered seemingly more than the actual economic return.

A supplementary report on Wuhan's sources of drinking water also represented a similar effort to comfort the public.[9] According to this report, the water was supplied by the Donghu Water Facility, at least until 1986. The municipal authority upgraded the facility and claimed that the results were fairly satisfactory. However, it was up to the provincial administration to decide whether the facility should be replaced. On September 30, 1986, the municipality published a code on the protection of drinking water sources. This code followed the establishment of a daily monitoring system based on new investment for water-quality monitoring devices. To secure what was expected to be the temporary operation of the Donghu Water Facility, an investment of 2 million extra yuan helped enable it to produce water that was 98 percent pure. Such figures were probably not very reassuring for residents, who were exposed to a 2 percent risk of drinking polluted water whenever they turned on the tap. However, the new sewage treatment facility was under construction to work with the half-finished sewage interception project.

Environmental Regulation in 1987

The 1988 *Wuhan Municipal Chronicle* continued the practice of presenting an overview report on the previous year's environmental protection performance, though still under the section title of Urban-Rural Infrastructure (no longer categorized with Environmental Protection). This section listed the achievements of 1987 and claimed that many goals had been fully accomplished. Its overview report also noted that Wuhan's environmental quality remained the same as in the previous year, lauding increasingly regularized environmental regulations. These achievements would also include certain levels of policy flexibility for lowest-ranked offices and environmental officers, consolidating an integrated management system combining regulation, research, and monitoring. However, as recognized in the *Chronicle,* most of these successful tasks involved collecting and managing pollution data and writing investigation or lab reports. For example, the Wuhan Bureau of Environmental Protection signed an agreement memorandum of advocating anti-pollution research collaboration with the US Environmental Protection Agency. This sounds like another pre-assigned assistance job from central authorities. The year-end achievements section included a list of twenty-four "clean and pollution-free factories" and eight "national nice environment factories."[10]

Another entry in the *Chronicle* discussed the collection of waste-discharge fees: the account registry work for the fees had reached 1,023 units fully charged, compared to 1,238 units registered city-wide in 1986. An increase of 253 registered accounts, compared to 492 accounts of the previous year, made a total collection of 25.75 million yuan versus 27.32 million yuan collected the year before. The underlying logic of this is consistent with the view that "green" taxes might reduce emissions and health damages and even enhance economic growth. Nonetheless, difficult trade-offs between economic development and environmental quality seem inevitable.[11] The decline in collected waste-discharge fees, from 1,238 units fully charged in 1986 to 1,023 in 1987, may imply some resistance to paying these municipally regulated fees, which could be considered a temporary or transitional payment. As such, citizens didn't pay such fees because they didn't feel legally required to do so, which wasn't the case for paying taxes. A briefing report seemingly provides a quick explanation:

> With strengthened auditing work for environmental project refunding, a total of ninety-two projects received approval with a total amount of 13.14 million yuan refunding for economic aid and improving environmental protection of these projects. Four million yuan were allotted to create a fund of small and low-interest loans, and this fund aimed to support local environmental business industries and pollution-reduction projects.

This report also noted that eutrophication had spread to almost all the urban lakes in Wuhan. Complaints about traffic noise still increased annually, yet industrial noise had seemingly been brought under control. Nearly all the major projects filed assessment reports, including some small projects though twenty-five reports were denied approval. Moreover, a notable shift of local emission regulation deserves more attention, and a second report manifested this change.[12] It noted that after 1984, ten city codes/bylaws were issued by the Wuhan municipal government and its environmental offices, another six were jointly administered by the latter and other government branches, and six internal directives had been produced in three years.

These figures sound quite impressive for the HEP, being a small newly established office. Besides these municipal reports on environmental protection, several internal document-profiling grey-paper publications compiled environmental protection laws and regulations by national, provincial, county, and municipal authorities. Celebrating World Environment Day on June 5, 1987, their anti-pollution campaigns, i.e., a series of state policy measures to contain pollution that were described as imperative in the government reports, aimed to promote environmental awareness among citizens via knowledge contests, photo exhibits, public forums on environmental law, and included a public competition of thesis-writing awards for environmental protection.[13] Serving an alternative platform of public intervention or governance engagement, these petition letters written by urban residents continued to protest incidents of pollution that the local authorities had ignored. In 1987, a total of 1,423 letters filing personal or group petitions raised environmental matters, 117 environmental motions or policy advice were proposed to the municipal committee of the Chinese People's Political Consultative Conference, and the Municipal People's Congress had addressed seventy-one pollution-related conflict appeals by local residents as civilian plaintiffs.

The shift of resorting to legal measures rather than relying on administrative approaches still had a long way to go, but a few promising signs had already appeared. For example, a fishery pollution incident occurred in the Hou Lak neighbourhood in Qiaokou in February–May 1987. A group of villager victims accused a group of factories of illegally discharging effluent. In dealing with the situation, the local district office of the HEP cited the National Environmental Protection Law and the Water Pollution Prevention and Control Law. In the end, it categorically mandated compensation of 151,000 yuan, collectively paid by eight effluent discharging factories led by the Wuhan Chemical Material Plant. In November 1987, the Wuhan Bureau of Environmental Protection had a joint consortium with the Wuhan branch of the China Society for Environmental Sciences. Attended by both environmental officers and scholars, it focused on local lake ecology and water source protection. Regarding how to combine the legalization job with academic discussions of correlated environmental policy research, we could consider this event one of the first experiments to integrate better environmental protection legalization with semi-public discourses among environmental officers and academics.

The rural-urban construction section in the 1988 *Wuhan Municipal Chronicle* frequently cited environmental assessment reports (EARs). EARs were also increasingly cited in the media and in government directives. These documents, offering potential solutions based on pollution evaluations and assessments, soon became mandatory paperwork for any builders acquiring permits for construction projects, based on the authorization of the updated National Environmental Protection Law. Because most environmental agencies were still under subprovincial bureaus of rural-urban construction, environmental protection business did not seem mature enough to be granted a stand-alone section in the *Wuhan Chronicle*. A 1988 report provided a few annual summary figures for the EARs.[14] For thirty-five large construction projects in Wuhan, all had filed and complied with their EARs. In their total investment of 1.092 billion yuan, 45.47 million yuan of environmental investment accounted for 4.16 percent of the total city-registered investment. A total of 271 small-medium construction projects had 4.74 percent of their total investment. Moreover, the document filing rate for the large projects was 100 percent, whereas it was 90 percent for the small-medium group.

Besides claiming that all the large construction projects had filed an EAR, the *Chronicle* report added a short review of early investigations into the locally high-profile effluent pollution case of the Jiangxia Gedian Chemical Factory in southern Wuchang. As noted in the city archival records, with a long list of external reviewers becoming involved with the factory's EAR procedures, new investment reached 5.5 million yuan for effluent treatment. The total anti-pollution investment reached 9.98 million and accounted for 6.2 percent of its total investment. Institutes that contributed to this EAR project included the Wuhan Environmental Research Institute as a lead reviewer, the Yangtze Water Resources Protection Institute, the Meteorology Application Institute under the Hubei Meteorological Service, the Wuhan Environmental Protection Monitoring Stations, the Hubei Institute of Hydrogeology and Engineering Geology, and the Wuhan Chemical Engineering Company. An expanding community of EAR participants seemingly created the foundation for the shift to applying legislative measures to supplement administrative approaches in enforcing regulation codes against industrial pollution.

Along with this shift, the reasoning of differentiated policy standards or implementation strategies sounded logical and pragmatic too. An extended survey by the municipal government agents divided urban Wuhan into four zones: the crowded residential zone; the water source protection zone; the tourist resort zone; and the education-cultural zone.[15] In the survey report, the most attention was paid to the first three zones, probably because they generated the most complaints about industrial pollution.

Notably, the densely populated residential zone in the Jianghan district of Hankou had the most cases of air and noise pollution. Living conditions in the Qiaokou district of Hankou were among the worst, with more than a dozen textile factories and sixty-five pollution sources that affected the safety of its drinking water. A total of 1,593 industrial waste emission sources was reidentified by surveying 425 industrial enterprises within the four zones or near them. Routine surveys in the zones helped in both the selection of priority agendas and the enforcement of regulation. The local environmental protection offices proposed to give more policy priority with increased monitoring routines to 70.4 percent of pollution sources and took various and practical measures for those polluters.

According to a 1988 report in the *Chronicle*, suburban Wuhan boasted eighty-two water facilities by the end of 1987.[16] Forty of them had been built during that year alone, with a total investment of 8.93 million yuan: 2.4 million came directly from the municipal budget and 6.53 million from local rural-urban communities, with more investment expected. A total of 568,300 residents relied on a daily drinking water supply of 23.13 million tons. These water facilities subsequently expanded their initial supply by a factor of fifteen, with nineteen times the original pipe extension and 493,300 new water bill accounts that had previously relied on wells, rivulets, or ponds. This stunning growth reflected a shift to tap water from earlier reliance on groundwater and surface water, which were becoming increasingly contaminated, an ironic deterioration given that Hubei is known for its water resources. The report on Wuhan's water facilities also stressed frugal investment and called more policy attention be paid to energy conservation. Most water pollution is produced by either household consumption or industrial manufacturing. Many entry items stressed new civil building codes made by environmental protection and industrial waste regulation offices in Wuhan, including streamlined office structures and processing procedures for better policy efficiency.

The records in the *Chronicle* show that local environmental agencies and communities jointly attempted to address pollution issues while integrating their resolutions with some technical and policy ingenuity. Environmental governance could be better built on public responsibility, government accountability, and inclusive community-based commitments without necessarily clashing with local development goals. A generally shared responsibility would also have implied a solid foundation for public governance that engaged all stakeholders, including local state institutions, civil communities, and individuals. Although the focus of this book is on industrial pollution and environmental governance at a regional scale, a similar analysis may be applied to other public policies or regulation issues. In 1987, Wuhan's environmental monitoring network celebrated its fully functioning capabilities, credited to a wide range of organizational support, including a long list of research institutes and universities.

According to the above report in *Wuhan Chronicle,* the deployment of the Wuhan Municipal Environmental Monitoring Network indicatively started in 1981; in that year, the facility officially finalized its pilot tests for both air quality and noise monitoring. In 1982, it finished the groundwater-quality monitoring network, followed by its precipitation monitoring network in 1983 and an air quality automatic monitoring network in 1984. In essence, Wuhan established a full-scale monitoring network for smoke-dust, noise, and wastewater pollution in three years. In 1987, this network uploaded 7,868 data entries for dust monitoring, 1,170 entries for urban noise, and 7,754 entries for surface water samples from two rivers (the Changjiang River and the Han River), eight urban river sections, and eleven lakes. Because of its outstanding contribution, Wuhan's central environmental monitoring station, a capstone for the monitoring network, received national recognition.[17] The station alone created and uploaded 64.6 percent of this monitoring network. During the sixth five-year plan, it finished its monitoring data support for sixteen research projects, eleven of which were accomplished with national or subnational monetary awards. The station profiled the annual reports of municipal environmental monitoring from 1984, with a first attempt to digitize its publications, an environmentally friendly move even by later standards.

The municipal environmental offices kept investigating violations of the newly issued environmental law and paid attention to water pollution within their administrative borders. Two serious cases came to light: the Xunshi River in Wuchang and Moshui Lake in Hanyang.[18] Some land zoning was planned for new college campuses near the Xunshi River. Both the river and Moshui Lake would require large-scale sewage interception projects, like the one for Dong Lake, whose first phase was completed in 1986. The Wuhan City Zoo (among the largest Chinese zoos) was on the shore of Moshui Lake, and a city-planning proposal had designated the lake as within the tourist resort zone. Household sewage and industrial effluent were the confirmative leading causes of eutrophication in the lake: about 63 percent of its urban effluent came from 130 factories, with the remaining 37 percent from household sewage.

After years of sporadic community protest and tough negotiations to settle compensation amounts for pollution-related damages, pollution reporting finally reached the city budget table. These cases might represent instances of passive or procrastinating responses against environmental protection pressures. Promoting positive cases of more proactive regulation measures also occurred to Wuhan's environmental practices. Reports on eco-farming attempted to popularize rural-agricultural environmental protection, such as at Huangjingqiao village in Wuchang county.[19] The village exemplified environmental awareness via eco-farming, through the judicious use of chemical fertilizers and pesticides, and by conducting less monoculture. The progressive awakening of environmental awareness appears to have been strongest in areas that had been affected by early industrial pollution.

A report of 1988 raised hopes regarding the revival of ecological diversity. After years of gradually returning some croplands to their original condition as lakes or wetlands, residents of the Chenhu area in Hanyang county (and the news media) witnessed the return of a growing number of migratory Oriental white storks. In previous years, the numbers of many overwintering migratory birds had diminished, so the return of the storks was seen as an auspicious message. When more than thirty of the beautiful birds were killed by poachers, a criminal investigation ensued. The restoration of biodiversity was still a dream, but it delivered a hopeful message with a positive branding image. One lesson might be that where environmental crises are concerned, taking action may not always bring instantaneous results. Fully reversing decades of environmental degradation takes years and demands resilience, a coordinated effort, and a long-term struggle by many stakeholders. Rather than invoking dystopian despair, motivating hopes for a brighter future and building solidarity and community-based action are more desirable than an economically or politically divided society mired in ideologically motivated confrontations.

Another good news story in the 1988 *Wuhan Municipal Chronicle* was the announcement of the third-phase construction of the Huangxiao River project.[20] Scheduled to begin in October 1989, the new road-tunnel sections would include another 1,168 metres of box culvert with an avenue on top, plus 5,400 metres of a new open waterway for rerouting and dredging. The project called for sectional bids from an army of construction companies, some of which belonged to the city civil construction group, such as the 1st and 3rd Municipal Engineering Companies and the 3rd Municipal Construction Company. Others were military construction companies and hydraulic engineering companies. The latter won most of their bids for the silted sections of the watercourse that needed excavation, while the military companies were also awarded some section contracts. Hankou's civil servants voluntarily or mandatorily undertook a few auxiliary construction works. The project was expected to finish by 1989. The report on the Huangxiao project could be correlated with the timing of another report updating issues on river clearing.[21] The *Chronicle*'s subsection on dam-dike construction included several reports on regularizing daily maintenance of watercourses to prevent them from breaching their banks or waterlogging urban areas during the flood season.

The key issues regarding pollution regulation would all point to the difficulty of finding accountable personnel or organizations and taking quick action, ideally before the water rose on rainy days. The Huangxiao River project illustrates that some policy problems for local waterbody repairing must invoke collective initiatives. The above reports on the Huangxiao River project and river clearing consolidate the argument that well-coordinated environmental protection measures can help create a more comprehensive foundation of environmental governance. It may not always function perfectly, but advocating for more policy consensus is always useful, especially when immediate and resolute action is necessary. For many locals, the principle of survival would always be the golden rule, the mandate for continuance, even in the most excruciating situations.

Consolidating Environmental Codes in 1988–90

The year 1988 came in the middle of the seventh five-year plan; there were more consolidating signs for effective regulation practices by Wuhan's municipal environmental protection offices. Again, the annual overview first presented in the section on environmental protection business was now a stand-alone section in the *Chronicle*. It listed three main achievements for environmental protection in 1988.[22] First, the Wuhan Bureau of Environmental Protection had created twelve dust-control zones, more than twice the original goal of five. The total dust-control areas now covered 52.66 percent of urban Wuhan. Second, rather than developing a list of only ten "clean and pollution-free factories," the municipal bureau had awarded that status to seventeen. Third, the original annual goal had been to finish regulating sixty factories with sixty-three pollution sources. Instead, this number had reached seventy-nine factories with eighty-two emission points. Environmental deterioration had been partially halted thanks to the intensified level of effective regulation. Environmental monitoring data revealed that the quality of both air and water showed signs of improvement, whereas urban noise levels had been curbed. The waste-discharge fees, collected from 1,200 accounts, reached 24 million yuan, with a refund of 19.98 million to 195 pollution-control projects. As the bureau had accomplished all of its administrative goals, the municipal authority recognized its excellent performance. Ninety-nine individuals were also rewarded for their remarkable contributions.

A new five-tranche categorization system was applied to all the large water bodies in Wuhan. The highest tranche would have the cleanest water. Fortunately, the Changjiang and Han Rivers belonged in this category. The Fu River was in the second tranche, Dong Lake was in the third, and the rest were in either the fourth or the fifth. The average noise level for Wuhan's districts dropped 2.5 decibels, still higher than the required noise control standards, except for the cultural-education zones. This improvement was attributed to the ease of locating the sources of the noise. High decibel levels resulting from commercial activities appeared easier to address after identifying the noisemakers. However, in Hankou, fines or levies had somewhat limited effects as the electrical feedback noise from mega-speakers led to even more complaints afterward.[23] It was challenging to argue with locals who were proud of their loud voices.

A report on air pollution presented optimism for the monitoring outcomes of 1988.[24] The Atmospheric Pollution Prevention and Control Law of the PRC was passed on September 5, 1987, and enacted on June 1, 1988.[25] Nonetheless, efforts to control industrial dust had begun to occupy a priority position in the working agendas of local environmental offices. A quick response to the enforcement problem from the HEP may refer to the establishment of the Environmental Protection Court at Qiaokou, under the approval of the Hubei Provincial Supreme Court.[26] Two new cases of effluent pollution occurred during 1988, both involving chemical discharges and accidental manufacturing malfunctions.[27] Some districts suffered major damage as a result, but neither case went to court, though the fines were paid. In another case, a proven violator, the Wuhan Distillery, refused to pay a meager fine of five thousand yuan. Eventually, it lost its case at the Environmental Protection Court and had to pay the total amount immediately without resorting to an appeal.

Another relevant event deserves some attention: In October 1988, Wuhan hosted an internal conference of eight municipal environmental protection bureau directors. This semi-public meeting invited seven municipal bureau directors from Ha'erbin, Dalian, Shenyang, Xi'an, Qingdao, Guangzhou, and Chongqing, besides Wuhan. All these cities fell into the "subprovincial" city category, and the event was the third of a conference series that was designed to promote regulation, intelligence sharing, and information exchange. The directors agreed to work together on six priority agendas, from reinforcing accountability management to expanding the environmental protection agency network and strengthening their personnel programs. These agendas seemed like a pragmatic way to coordinate regional environmental protection work, so that Wuhan's pollution regulation performance would look consistent with its reports.[28] In the following year, the 1989 *Wuhan Municipal Chronicle* assessed a list of 1988 working agendas as accomplished at the level of municipal bureau director.[29]

These reports of pollution-regulating assessment stressed strengthening local leadership, further publishing the first annual report of quantitative evaluation, and the effluent-discharge licence system. On the basis of the performance evaluation review standards designed by the provincial government and mandated by the National Environmental Protection Committee in the State Council of the PRC, the Wuhan government well achieved its pollution regulation goals. In October 1989, Wuhan hosted the second training program for environmental archivists, as deputized by the National Environmental Protection Ministry.[30] A total of forty officer trainees came from eastern, northeastern, and southern China. Such training programs had started a few years earlier. In 1986–87, the Wuhan Municipal Bureau collaborated with the School of Law at Wuhan University, whose acceleration program on National Environmental Protection Law had issued 176 certificates after one year of study. In 1989, in two similar training programs designed for environmental business management, fifty-six and forty-five students (presumably public servants) graduated with professional diplomas. An education centre for teaching environmental protection was established in 1989 by the municipal government to provide vocational or paralegal training for environmental protection. More positive news came from these popularizing campaigns.

According to the 1989 report, the performance evaluation for waste regulation and environmental protection in Wuhan received a decent score: for comparison, out of a standardized index, Wuhan was reported to have gained a total score of 56.6 out of a full/highest achievement index of 100, across its twenty projects that were registered in the National Environmental Protection Committee index. Wuhan's environmental protection score was ranked at eighth among thirty-two cities of policy priority selected by Beijing. The committee designated most major cities in China to undergo an annual quantitative evaluation on comprehensive urban pollution management and environmental governance. Its routine annual reports tracked the research projects of local environmental science institutes.[31] As in earlier years, the *Wuhan Municipal Chronicle* of 1990 published a review of two environmental research institutes in Wuhan: one service centre of environmental protection technology funded by the municipal government and one institute established in 1978 under the joint supervision of the Ministry of Water Resources and Electric Power and the National Bureau of Environmental Protection.

However, one of the most depressing news items in the 1990 *Chronicle* was an environmental research report that predicted the extinction of the Chinese river dolphin, also known as the Yangtze River dolphin or *baiji* (meaning white-fin in Chinese) dolphin.[32] An earlier report had confirmed that the dolphin was last seen in the winter of 1987, when sailors spotted it upstream from Wuhan. Afterward, there was little solid evidence to show that it survived in the wild. From the 1950s, a few of these extremely rare freshwater dolphins were sheltered in seventeen river sections between Hubei and Anhui, some of which provided silt-encircled asylums. Others were kept in a massive tank at the research lab of the Wuhan Institute of Hydrobiology, which is part of the Chinese Academy of Science. Local hydrobiology scientists estimated that fewer than three hundred dolphins were left in the open water by 1989, compared to an estimate of approximately a thousand two decades earlier. The dolphin has been nicknamed the "giant panda of the water," but it does not share the success story of its terrestrial counterpart. Despite all the efforts, it has not been pulled back from functional extinction. The last scientist-sheltered dolphin, named Qiqi, died at Wuhan in 2002, and as recently as 2018, locals were still looking for proof that the species survived in the wild.[33] The factors contributing to its demise, publically confirmed in the official media, include industrial pollution, overfishing, shrinking habitat related to human activity, underwater noise due to increasing river traffic, fisheries construction by riverbanks, and the building of dams and bridges in the river-lake system.

Routine summary reports for 1989 indicated that most environmental regulation and policy implementation was regularized as during the previous years of the seventh five-year plan. An annual review report also noted that economic development had slowed during the year. However, the collection of waste discharge fees on the city scale sounded satisfactory, at 26 million yuan, with a count of 8 million yuan for ninety-nine environmental protection projects applied by eighty-four enterprises. The slowdown in economic growth was not limited to Wuhan. From 1989, Western countries imposed a series of economic sanctions on China after the massacre at Tiananmen Square. Labelled an authoritarian regime and anti-democratic, the Beijing government was severely criticized. However, Wuhan was far enough from Beijing, both geographically and culturally, to resist external pressures and focus more introspectively on its local issues.[34] This status would make its policy challenges different from those of other major cities.

An annual review report for 1989 addressed pollution produced by village-owned or township-owned factories.[35] It called for more policy attention and stricter regulation of new waste emitters in rural and suburban Wuhan. Most counties increasingly detected acid rain. Despite some advances in regulation and positive pollution control measures, emissions from numerous small factories still threatened the rural ecology.[36] In the urban-rural space of Wuhan, a policy experiment with issuing effluent discharge licences targeted ten factories near Dong Lake, Moshui Lake, and Nan Lake.[37] It issued two types of licence: the standard version was granted to factories that had not reached their quota of industrial effluent in previous years; the temporary version was given to factories that exceeded their discharge quota. Such a small-scale experiment probably garnered limited attention. Instead, local resident communities paid more attention to the massive rainfall of 1989.[38] The summer was relatively cool and the autumn rains were heavier than usual, sparking concerns for cotton and rice production, which was expected to be much lower than average. Nonetheless, 1989 looked like a typical year of little significance for Wuhan.

In 1990, a relatively chilly and rainy spring, with precipitation concentrated in February and April–June, was followed by an early and extremely hot summer and a rare dry autumn.[39] Local scientists had already noted a pattern of climate warming: from 1986 to 1990, a temperature increase of 0.1 to 0.3°C was recorded in most parts of Wuhan, except in its hilly Huangpi county, which was at a higher elevation than the urban districts. In this situation, a short novel published in 1991 by a local writer, Li Ci, became popular, and its title, roughly translated as "hot or cold, it is good to live," quickly became a trending expression among people in Wuhan.[40] The novel concentrates on daily life in Hankou, vividly depicting both excruciating and joyful moments in a locale where public and private spaces are blurred. Its characters casually chatted about the 1990 invasion of Kuwait. The year 1990 created memories of mixing order and chaos in such a crowded urban space, along with some reality-based narratives in this well-known novel. One can relate its artistic reflections to Wuhan's space-time evolution as follows: The urban population of Wuhan increased by 1.94 percent annually or 10.45 percent in four years, with a net growth of 633,600 residents, reaching 6.698 million in 1989, up from 6.064 million in 1986.[41] The total population of Wuhan had reached 13.64 million by 2021.

Apart from the hot summer of 1990, the weather looked ordinary. The winter was not as cold as in the previous year, except for a few more rainstorms and occasional urban waterlogging; overall, no significant weather events seemingly occurred in Wuhan. However, environmental governance at the national level continued to evolve. On December 26, 1989, the first

National Environmental Protection Law was passed. The seventh five-year plan thus marked the first enactment year of this law, formally initiating its full legalization across the PRC. Many local narratives delivered a positive sense of optimism yet with a pragmatic "whatever works" strategy. The mission of popularizing environmental protection became a priority item in the *Chronicle*'s overview reports for 1990.[42] Such policy reports routinely summarized annual environmental governance lessons, with some negative news from previous years.[43] Another new industrial pollution incident was exposed.[44] Reports listed five additional "clean and pollution-free factories," for a total of seventy-eight factories as good illustrations of environmental protection in the city. A few received national awards for distributing their newly updated positive experience. The *Chronicle* noted some mixed results: The urban air quality had reached national alert level two; pollution was seemingly curbed in a few urban water bodies; and noise pollution remained much higher than the national standard. With rising cases of dust-related lung disease, even stricter regulations were proposed to apply to more workplaces.[45]

These reports in the *Chronicle* maintained an increasing call for more investment in many projects of overdue pollution control. The 1990 annual waste discharge fees collected 29.50 million yuan, and the refund policy had switched to an environmental project loan system as another experiment at the municipal level. Meanwhile, the municipal government requested that new funds be designated to improve the policy efficiency of pollution control.[46] A new account registry prepared work-to-issue effluent discharge licences. A total of 471 factories filed their applications for more policy experiments.[47] This effluent discharge account registry system seemingly aimed to open a new line of funding. An update report on the Dong Lake sewage interception project showed that it was still under way.[48] By 1990, it had installed a sewage treatment plant with auxiliary pump stations, as well as twelve of the twenty-four required kilometres of main pipes. Completing the project for districts surrounding Dong Lake, thirty-three square kilometres of which were designated a tourist-resort zone, consumed the entire period of the seventh five-year plan.

A report in the 1991 *Chronicle* conveyed some good news: the creation of Tian'ezhou Yangtze Dolphin National Nature Reserve, a protected zone whose initial purpose was to aid in saving the Yangtze dolphin.[49]

Once the natural habitat of the dolphin, the reserve was in a silted cul-de-sac section of the Jingjiang section of the Yangtze River, in Shishou county, Jingzhou city. In March 1990, it was officially opened, though it focused on another endangered river mammal, the Yangtze finless porpoise, known locally as the river piglet, a cousin of the dolphin. Three were captured in the wild and released in the reserve, where it was hoped that they would breed. On February 5, 2021, the porpoise was officially listed as a Class 1 protected species (an index the UN also recognizes), upgraded from its Class 2 status of almost thirty years and after more than two years of proposal reviewing.[50] Today, more than a hundred porpoises flourish in the reserve, a potentially hopeful sign for the future.

The events discussed above echo a set of protracted campaigns to address lasting environmental challenges in Wuhan. An increasing number of communities actively engaged in the struggle for better civil governance. The environmental offices in Wuhan understood that politicized environmental movements created a potentially perilous and artificially amplified risk of polarizing communities rather than drawing them together in the fight against pollution. Some might argue that negative media coverage would neither resolve this critical problem nor help enforce fast-track legalization of environmental protection or relevant state-policy measures of any kind. It is essential to engage state agencies and local communities through proactive policy lessons for knowledge promulgation and mobilization. Like Wuhan, many young cities in China have struggled to establish their own environmental governance, which seemingly emerged formally as recently as the late 1980s.

Conclusion

In 1986, the *Wuhan Municipal Chronicle* began to reliably publish reports on local efforts to address pollution and protect the environment. Cross-referenced with the records discussed in earlier chapters, this material covers more environmental policy experiments during the seventh five-year plan of the PRC, which began in 1986 and ended in 1990. The *Chronicle* routinely published review reports on environmental matters, revealing to some extent that local environmental protection offices had an oft-renewed sense of emergency to advocate for community-based

FIGURE 13 *Top:* A researcher feeds Qiqi, the captive river dolphin. *Bottom:* Training a Yangtze finless porpoise at the Baiji Dolphin House in Wuhan. Because its slightly curved lips resemble those of a smiling person, the porpoise is dubbed the "smiling angel of the Yangtze River." | *Wuhan Municipal Gazetteer* (Wuhan: Wuhan Municipal Chronicle Office, Wuhan Publishing House, 1986); *Xinhua Daily,* December 18, 2023.

efforts. At the same time, the semi-public reports emphasized positive policy lessons rather than raising the profile of negative cases. The material in the *Chronicle* provides an instrument to engage with local environmental governance. Carrying collective memories and policy-effect evidence of significant events, it preserves details regarding progress in the struggle to protect the natural environment under Wuhan's skies and documents the gradual but steady formation of Wuhan's agenda of environmental governance.

Epilogue

Grounded in a careful reading of archival material, this book aims to make an original contribution to the literature on the environmental history of China. Confining its scope to the province of Hubei during the transitional period of the 1970s and 1980s, it focuses on official and quasi-official records. Initially, regulatory policy largely failed to control industrial waste emissions, but reports on cases of industrial pollution in Hubei and its capital, Wuhan, show that the provincial and municipal administrations resourcefully adapted to national policy guidance. These archival findings help to conceptualize the ongoing progress of industrial pollution regulation in China. Both official and public pressures on local environmental protection state-agencies are part of this important narrative arc and are deeply embedded in China's historical experience.

Because of the lack of formally recognized avenues for public advocacy of governing policies in China's administrative system, I focus on the paths by which public pressure produced political will within a fragmented bureaucracy. My discussion does not extend beyond Hubei, and it is probable that environmental deterioration would be handled differently in other provinces. Some archival sources have confirmed an ongoing information exchange among various groups in Hubei. As to the roles of the HEP officials, they were primarily witnesses and conflict-solution providers in crises. They were also active members of the community, sharing social spaces with networks of families and friends. To deny their multiple identities or ridicule their inability to act more effectively is to take them out of their historical context. I prefer to regard them as presumably honest witnesses who had stakeholder interests.

My policy analysis rests partially on juxtaposing the 1970s and 1980s with the present and the late imperial era, as I examine the official side of pressures and activities and, to a good degree, the social or public sides as well. As they had been before the late-imperial period, the PRC provincial and subprovincial officials continued to be on the front lines; they provided tentative policy solutions to ecological crises in an effort to keep the peace among disparate interest groups. Emphasizing their pivotal role in governance systems is vital to future examination of relevant policy matters. The introduction of receptivity into Chinese scholarship, as briefly mentioned earlier, also needs to be stressed here. Highlighting an explicit connection with how local communities and state agents fostered environmental governance, this book helps to illustrate the trajectories of environmental policy, specifically how official environmental protection agencies in Hubei were established, functioned, or interacted over time, always in the context of some inherent institutional weaknesses.

In examining this subject, I stress neutral interpretations that do not endorse binary narratives, such as good agriculture versus bad industry, idyllic countryside versus urban chaos, or ignorant government agencies versus righteous citizen protesters, all of which can be easily connected to inadvertent presentism and anachronism, the *post hoc ergo propter hoc* fallacy. This fallacy holds that because two events occur at about the same time, the first event must have caused the second one. In my research, I consider the first event to be the damage caused by industrial pollution in Hubei, and the second event to be the province of Hubei's poor record on environmental governance. Interpretations that are biased with populist views or that look down from the moral high ground of the present day are ultimately not useful. The deeper one delves into this subject matter, the more one recognizes that the environmental policies existed within rich and complicated socio-cultural-historical contexts.

In the eyes of subnational officials in Wuhan, many sensitive and seemingly sentimental issues regarding environmental fairness conflicted with the economic development agenda of the state. Consciously or unwittingly, local decision makers, including both government officers and state-owned factory managers, applied economic reasoning toward state policies when dealing with rising public pressure to address industrial pollution. Legalization of environmental protection gradually started to improve industrial waste emission, with institutional constraints at national and subnational levels. However, some environmental protection measures materialized only in the late 1970s, and one should not employ concepts or ideas that have emerged more recently to make anachronistic interpretations.

Here I need to reiterate a few potential directions for further research, while deliberately keeping in mind the need to neutrally interpret the right to pollute in Hubei. First of all, my inquiry has been mainly confined to that province, a remarkable case of presenting rich features of subnational environmental governance; also, more of the relevant evidence came from cities rather than villages. Second, economic conflicts between the agricultural and the industrial manufacturing sectors can be interpreted as one of the most intensified sectorial tensions that occurred during China's national experiments with industrialization. As locally observed and reported first, poorly regulated industrial pollution was largely responsible for environmental decline. Last, when early signs of ecological damage emerged from the 1960s, a series of policies concerning the enforcement of industrial waste control had already penetrated China's vast inland areas. In Hubei, a series of industrial pollution regulations and rural energy and cyclical economy policies had followed the underlying policy reasoning, at least from the early 1970s, continuing to the late 1980s.

Most of Hubei's environmental protection officers admitted that their work fell short of public expectations and could perhaps even be seen as a total failure. There is no point in applying a post hoc judgment to their performance since even the ad hoc rules under which they operated seemed to have been vaguely defined. In their defence, it is unrealistic to expect that they should have been more willing to override state policies and other initiatives that prioritized industrial development, or indeed that they were capable of doing so. Many environmental initiatives were overshadowed by the catastrophic Great Leap Forward (1958–62) and the Cultural Revolution (1966–76). Political transitions significantly complicated the practical challenges of local environmental governance. China's leaders aimed for a swift nationwide industrialization, but this endeavour tragically devolved into a series of continuous ideological confusions and internal political struggles. Later economic incentives helped promote environmental policy efforts and mobilize the masses for more proactive measures against industrial pollution. It is easy to condemn Chinese officialdom for its lack of environmental awareness, but this is not the point here, as I focus on sharing more progressive and reality-oriented initiatives serving to address the environmental issues at hand.

Economists in the PRC, including its environmental economists and industrial policy specialists, would fall into a subcategory of intellectuals. Leftist scholars commonly surveyed in the intellectual history studies of China, extending from economists and geographers to sociologists, labelled these presumably well-informed individuals as traditional "gadfly writers" or "literary figures." The term "gadfly writer" can also narrowly refer to the policy and social and intellectual history frameworks of Chinese history, as explored in classic works by Western historians of modern China, such as Goldman and Gordon (2000), and Hamrin and Cheek (2023). I would argue that more cross-disciplinary dialogues in the future should reflect on why, in particular, enviro-economists have had a greater role than other intellectual members of the Chinese Academy of Social Sciences, including ecologists and natural scientists. It seems that the important role of economists in local higher education systems, in the CCP political system, and also deeply embedded in the local and regional institutional framework (bureaucracy particularly) is critical.

Some archival evidence confirms the existence of Hubei's environmental protection policies in controlling large-scale industrial pollution, despite their disappointing effects. I doubt that a personal moral code could sufficiently incentivize county and municipal government officials, industrial officers, and managers of state-owned enterprises to take voluntary, pro bono actions for coupling short-term zero-return technological investment with pollution reduction. Relying mainly on moral incentives, as illustrated by the level of corruption at Hanyeping, did not have a solid governance foundation to sustainably coordinate economic factors in terms of the social contract or political arrangement between the general public and individuals. Nothing in the archival sources for Hubei indicates that any environmental NGO existed in the province during the 1970s and 1980s, and indeed their rise occurred at a much later date in the PRC. It would be best not to interpret these late arrivals as a curse or a blessing but simply as beyond the scope of this book.

Here is a question worthy of consideration: How does one fairly and continually evaluate Hubei's provincial system of environmental protection bureaucracy within its local governance settings? My analysis highlights accountability issues and bureaucratic weaknesses among provincially intertwined conditions to explain the restrictions of enforcing environmental regulation in Hubei. The archival material reveals how local authorities adapted some protection policy measures to other imperative tasks. Gradually and arduously, an environmental governance mechanism developed in Hubei and particularly in Wuhan. Regarding the remediation activities documented by environmental protection agencies, I would contend that certain early policy efforts should be viewed positively as a good foundation for progressive policies and thus more effective regulation experiments, as executed from the lower levels up to the higher bureaucracy and eventually a much wider scope.

Regarding the beginning of environmental governance in the PRC, I suggest that awareness of pollution problems materialized early, even by the late 1950s. Industrial pollution particularly damaged water resources in the 1960s, but policy measures were not in place until the 1970s, when local authorities began to respond, especially to the unrest in rural fisheries and urban areas. Many provincial agencies sprang up; directives from Beijing followed later; the notion of fragmented bureaucracy plays into these pollution damage negotiations, with an emphasis on policy formulation centred in Beijing but relying heavily on decentralized implementation. This indicatively fragmented bureaucracy is especially illustrated in Hubei and Wuhan. The gradual and sometimes annoyingly belated evolution of government branches dated from the 1970s or even earlier, with a series of politically rendered but economically reasoned state-led campaigns.

A series of factors hampered early efforts to deal with pollution: bureaucratic conflict, inter-agency interests, underfunding, understaffing, and many practical difficulties in imposing regulations. On the establishment of Hubei's environmental agencies, this study underlines local resistance to central mandates and inducements, though certain mass movements did generate some initiatives. The biogas movement did not begin as an environmental concern but may have later developed into one. Compared to the Sanfei movement, it looked somewhat successful in the era of chemical fertilizers. Early state policy efforts for containing industrial pollution faded after the reform era started to disincentivize collective projects concerning environmental pollution regulation. Some practices of pricing the right to pollute evolved in Hubei, while the practical recognition of the right potentially compromised environmental governance in Wuhan.

My inquiries highlight the dynamics between central and provincial institutions concerning the implementation of environmental protection schemes and pollution regulation codes that were centrally mandated but locally operated. The key entity of this book, the Hubei Provincial Bureau of Environmental Protection (HEP), incorporated local monitoring scattered across Hubei that served at the lower rungs of the environmental agency system, while playing the role of liaison agency with many affected players and societies. Soon after Hubei's provincial environmental protection agency was unofficially pre-established in 1973, with assistance from research academies and universities, more regional environmental monitoring stations emerged, operating under the higher guidance. Some early environmental agencies were more reliant on their superintendent offices in terms of funding and personnel support, therefore being essentially still managed within the same administrative framework of Hubei or Wuhan.

A small group of scholars, mostly economists, realized the direness of the environmental situation from the 1970s onward. Their papers affected how the state managed the challenge, and their debates help to map the reasoning behind its policy response to the problem. Multidimensional outlooks helped examine evolving policies. As both economic and political initiatives were weak when it came to environmental protection, few policy options were left to deter pollution effectively. Rather than relying on utilitarian goals or administrative measures, a very limited set of policy enforcement instruments was employed to regulate rampant industrial pollution. Although every resident of Hubei could potentially become a victim of environmental disaster, identifying polluters and awarding appropriate compensation to individual victims posed another implementation issue: in these very large groups, the categories of polluter and victim would easily overlap, and some locals were considered to be both polluters and victims. This conundrum was particularly prevalent in cases of air and effluent pollution.

China's early economic discussions on environmental issues indicate that the bureaucracy had little interest in regulatory mechanisms, potentially ignoring reflections per regulation standards or enforcement measures. Early national guidance codes for pollution governance looked comprehensive, but they also left plenty of room for subjective judgments among the newly designated state agencies for environmental protection. Archival evidence helps reveal that provincial and sub-provincial environmental officers consistently committed to warning higher authorities of the increasing failures in regulation. My archival findings enrich interdisciplinary understandings of the regional environmental evolutions in Hubei and Wuhan. In fact, given that public access to the archival records is limited to the period before the mid-1980s, this book aims to expand beyond the boundary of what is possible for historical inquiry in China during the transitional period of the 1970s and 1980s.

State agents did not lack awareness of the many escalating environmental problems. My archival findings confirm the significance of subnational or provincial-level implementation of environmental protection policy via inter-bureaucratic negotiation, particularly concerning how Hubei officials established an administrative structure for their industrial pollution regulation work, as well as the predominately economistic approach to dealing with regulation problems and the early emergence of quantitively pricing the right to pollute as an adaptive solution for environmental protection policy implementation. Continual struggles for more funding and personnel management for public services or goods for environmental protection led Hubei's government officials to create more robust environmental governance and policy effects.

Two facts are relevant: First, Hubei's environmental officers had been resisting interference by either the central agencies or outsiders, rather than making those central directives and ideological issues as dominated in the national mass movements. Second, environmental offices had also resorted to public opinion pressure to acquire more funding during the 1970s and 1980s. Nonetheless, many environmental officers in Hubei and Wuhan still tried to apply industrial waste pollution regulations. They acted like diligent intermediaries between dissenting voices and deployed measures other than those that encouraged confrontation. Moreover, certain levels of environmental policy continuities and evolving adoption significantly featured their reality-based practices. These early policy experiments foreground some implementation features for regularizing and eventually institutionalizing environmental governance.

Based on archival evidence, this book provides policy references to some core theses and theoretical contentions on China's environmental studies by both Chinese and Western scholars. It aims to incorporate more dimensions in examining local environmental problems and development barriers in China, and to engage readers who are interested in China's modern history, environmental history, and policy studies. The abundance of archival records has made it possible for me to take a historical approach to understand the environmental policy of Hubei. The book contributes to further historical understanding of environmental governance at the provincial and subprovincial levels during a period of transition, as reflected in Hubei and Wuhan during the 1970s and 1980s.

Notes

Chapter 1: Mapping Hubei

1 This observation is based on the updated figures of 2016. National Bureau of Statistics, *Chinese Yearbook* (Beijing: National Bureau of Statistics, 2017), Table 2–6 and Table 3–9.
2 Peter Ho, ed., *Developmental Dilemmas: Land Reform and Institutional Change in China* (London: Routledge, 2005).
3 Judith Shapiro, *China's Environmental Challenges* (Malden, MA: Polity Press, 2012).
4 Philip C. Huang, "'Public Sphere'/'Civil Society' in China? The Third Realm between State and Society," *Modern China* 192, 2 (1993): 224–26.
5 Piper Gaubatz, "New Public Space in Urban China." *China Perspectives*, 4 (2008): 73–84. See also Gu Xin, "Plural Institutionalism and the Emergence of Intellectual Public Spaces in Contemporary China: Four Relational Patterns and Four Organizational Forms," *Journal of Contemporary China* 7, 18 (1998): 271–301.
6 J. Donald Hughes, *What Is Environmental History?* (Cambridge: Polity, 2006), 124–25.
7 Elizabeth J. Perry, *Anyuan: Mining China's Revolutionary Tradition* (Berkeley: University of California Press, 2012).
8 Mark Elvin, "The Environmental History of China: An Agenda of Ideas," *Asian Studies Review* 14, 2 (1990): 39–53. See also Bao, Maohong, "Environmental History in China," *Environment and History* 10, 4 (2004): 475–99.
9 Maohong Bao. "The Evolution of Environmental Policy and Its Impact in the People's Republic of China." *Conservation and Society* 4, 1 (2006): 36–54.
10 Wenhui Hou. "Reflections on Chinese Traditional Ideas of Nature." *Environmental History* 2, 4 (1997): 482–93. Also see Martin Melosi. *Effluent America: Cities, Industry, Energy, and the Environment* (Pittsburgh, Pa: University of Pittsburgh Press, 2000).

Worster, Donald. *Wealth of Nature: Environmental History and the Ecological Imagination* (New York, Oxford: Oxford University Press, 1994).
11 "Introduction" in *Sediments of Time: Environment and Society in Chinese History*, ed. Mark Elvin and Cuirong Liu (Cambridge: Cambridge University Press, 1998), 13–18.
12 My writings may also refer these continuing threads taken by William Cronon, *Nature's Metropolis: Chicago and the Great West* (New York: W.W. Norton, 2009), besides the case studies by Martin V. Melosi, *Effluent America: Cities, Industry, Energy, and the Environment* (Pittsburgh: University of Pittsburgh Press, 2000).
13 Jiayan Zhang, *Coping with Calamity: Environmental Change and Peasant Response in Central China, 1736–1949* (Vancouver: UBC Press, 2014).
14 Rachel E. Stern, *Environmental Litigation in China: A Study in Political Ambivalence* (Cambridge: Cambridge University Press, 2013).
15 Tseming Yang, "Mysteries, Myths, and Misunderstandings," *Environmental Forum* 33, 2 (2016): 36–42.
16 Yang, "Mysteries, Myths, and Misunderstandings," 41–42.
17 Paul A. Barresi, "The Chinese Legal Tradition as a Cultural Constraint on the Westernization of Chinese Environmental Law and Policy: Toward a Chinese Environmental Law and Policy Regime with More Chinese Characteristics," *Pace Environmental Law Review* 30, 3 (2013): 1156–21.
18 Yuhong Zhao, "Environmental Dispute Resolution in China," *Journal of Environmental Law* 16, 2 (2004): 157–92.
19 Faure, Michael G., and Jing Liu. "Compensation for Environmental Damage in China: Theory and Practice." *Pace Environmental Law Review* 31, 1 (2014): 240–321.
20 Adam J. Moser and Tseming Yang, "Environmental Tort Litigation in China," *Environmental Law Reporter* 41, 10 (2011): 10895–901.
21 Vaclav Smil, *The Bad Earth: Environmental Degradation in China* (Armonk, NY/London: M.E. Sharpe/Zed Press, 1984).
22 Smil, *The Bad Earth*, 199.
23 Bryan Tilt, *The Struggle for Sustainability in Rural China: Environmental Values and Civil Society* (New York: Columbia University Press, 2010).
24 Elizabeth Economy, *The River Runs Black: The Environmental Challenge to China's Future* (Ithaca: Cornell University Press, 2010); Elizabeth Economy, "Environmental Governance: The Emerging Economic Dimension," *Environmental Politics* 15, 2 (2006): 171–89.
25 Abigail R. Jahiel, "The Organization of Environmental Protection in China," *China Quarterly* 156 (1998): 757–87.
26 Tianguang Meng, Jennifer Pan, and Ping Yang. "Conditional Receptivity to Citizen Participation: Evidence from a Survey Experiment in China." *Comparative Political Studies* 50, 4 (2014): 399–433.
27 Sam Geall, *China and the Environment: The Green Revolution* (London: NBN International, 2013).
28 Richard Louis Edmonds, "The Environment in the People's Republic of China 50 Years On," *China Quarterly* 159 (1999): 640–49; Richard Louis Edmonds, "Studies on

China's Environment," *China Quarterly* 156 (1998): 725–32; Economy, "Environmental Governance."
29 Shapiro, *China's Environmental Challenges*. For further discussion at the national level, see Bruce Gilley, "Legitimacy and Institutional Change The Case of China," *Comparative Political Studies* 41, 3 (2012), 259–84.
30 Andrew Mertha, *China's Water Warriors: Citizen Action and Policy Change* (Ithaca: Cornell University Press, 2008).
31 Emily T. Yeh, "The Politics of Conservation in Contemporary Rural China," *Journal of Peasant Studies* 40, 6 (2013): 1165–88; see also Emily T. Yeh, Kevin J. O'Brien, and Jingzhong Ye, "Rural Politics in Contemporary China," *Journal of Peasant Studies* 40, 6 (2013): 915–28.
32 Yeh, "The Politics of Conservation in Contemporary Rural China," 915.
33 Judith Shapiro, *Mao's War against Nature: Politics and the Environment in Revolutionary China* (Cambridge, MA: Cambridge University Press, 2001).
34 "Geography Section," *Hubei Shengzhi (HBSZ)* (Hubei: Hubei Renmin, 1997), 1–3.
35 Annual flood records for Jingzhou. "Jingzhou Diquzhi," *HBSZ* (1997) 84–91.
36 Xu Xinchuang et al., "Hubeisheng jin 500 nian quyu ganshi xulie chongjian jiqi bijiao fenxi" [Hube's 500 years of Dry versu Wet Weather Re-modeling and Comparative Analysis], *Geography Research* 29, 6 (2010): 1045–55.
37 "Geography," *HBSZ*, 1184–87 (Hubei Gazetteer Office, Wuhan: Hubei Renmin Publishing, 1990).
38 Zhang, *Coping with Calamity*, Figure 2.
39 Qixiang Tang, "Yumeng Yu Yumeng Ze" [Yunmengg and Yumeng Mash], *Fudan Journal (Social Sciences Edition)* 1 (1980): 1–11. For more on climate changes in ancient China, see Mark Elvin, 1998.
40 *Wuhan Shizhi (WHSZ)* (Wuhan: Wuhan Publishing House, 1990), 83–85, 32, 54, 69; Yunmeng Nianjiang, 2013, 55 (Wuhan: Hubei Renmin, 2013).
41 Jianming Zhang provides a comprehensive analysis on the dike-farm in the Han River–Dongting Lake area in Qing China, see Zhang, "Qingdai Jianghan – Dongtinghuqu Diyuan Nongtian De Fazhan Jiqi Zonghe Kaocha" [a Comprehensive Analysis on the Dyke-Farm in the Han River-Dongting Lake Area in Qing-China], *Chinese Aggriculture History* 2 (1987): 72–88. Also, *HBSZ*, 205–8 (Hubei Gazetter Office, Wuhan: Hubei Renmin Publishing, 1990).
42 Wuhan Shizi-Comprehensive (Wuhan: Wuhan Publishing House, 1990), the names of districts and counties of Wuhan, 100–10.
43 Jiayan Zhang, 2014. See also Jin-liang Huang, "Jin 500 Nian Jianghan Pingyuan Huqu Tudi Kaifa De Lishi Fanshi" [Historical Thought on Land Exploitation in Recent 500 Years in Jianghan Plain], *Journal of Central China Normal University (Natural Science Edition)* 35, 4 (2001): 485–88; and Sheng-sheng Gong, "Historical Variation and Sustainable Utilization of the Jianghan-Dongting Plain's Wetland," *Resources and Environment in the Yangtze Basin* 11, 6 (2002): 569–74.
44 "Water Resources," *HBSZ* (1990), 369–80.

45 Jason K. Levy, "Multiple Criteria Decision Making and Decision Support Systems for Flood Risk Management," *Stochastic Environmental Research and Risk Assessment* 19, 6 (2005): 438–47.
46 HPA. The reports of budget request for continual infrastructure consolidation: SZ113-2-285-002-6 in 1964, and SZ139-6-0358-004-5, 0550-001, 0486-001 in 1971–74.
47 David Allen Pietz, *The Yellow River: The Problem of Water in Modern China* (Cambridge, MA: Harvard University Press, 2015).
48 See the Introduction, *HBNJ* (1997), 2–3.
49 See the Introduction, *HBNJ* (1989), 44. "Geography," *HBSZ* (1990) 2.
50 "Water Resources," *HBSZ*, 341–69, 423–25.
51 "Water Resources," *HBSZ*, 546–49.
52 Qinzi Rao. "Hubeisheng Hubo Diaocha" [A Survey of Hubei Lakes], *Chinese Science Bulletin* 10 (1954): 71–83. This survey covered 591 lakes, as documented also by Qingquan Xue. "Zhongguokexueyuan Zucheng Hubo Diaochadui Qu Hubeisheng Yanjiu Yulei Yangzhi Wenti" [Fishery Questions of Hubei Investigated by the Lake Team Sent by the China Institute of Science], *Chinese Science Bulletin* 7 (1953): 97–98.
53 Hongbin Deng, et al. "Jin 50 Nian Lai Jianghan Hu Qun Shuiyu Yanhua Dingliang Yanjiu" [Quantitative Research on Hanjiang Lake Groups in 50 Years], *Resources and Environment in the Yangtze Basin* 15, 2 (2006): 244–48.
54 One mu of area unit in China equals to 1/15 of a hectare, or nearly 666.67 squared metres. A thousand mu equals 0.667 square kilometres. Refer to the Overview section, *HBNJ* (2015), 24. Local authorities accept a record of about 800 lakes above 100 mu today. An early record was a figure of 320 lakes over 3 square kilometres. Also refer to the Overview section, *HBNJ* (1989), 51–52, 262.
55 Some of the material in this section was originally published in Yun Liu, "Revisiting Hanyeping Company (1889–1908): A Case Study of China's Early Industrialisation and Corporate History," *Business History* 52, 1 (2010): 62–73.
56 Chen, Xulu, and et al., eds., *Hanyeping Gongsi* (Sheng Xuanhuai Archive Selection IV) (Shanghai: Shanghai Classics Publishing, 2015), Volume I:3, I:24. This published archival collection (with its pre-print editions) includes thousands of items, from telegraphs and mail to secret reports and financial records, held at the First Historical Archives (HYP), Beijing.
57 Xulu et al., *Hanyeping Gongsi*, I, 2; II, 672.
58 The Chinese Ministry of Metallurgy, *The Chinese Iron and Steel Industry Yearbook* (1985–2008), Beijing: Metallurgical Industry Press, 2009.
59 Chen and et al., *Hanyeping Gongsi*, 1: 6–8.
60 Tianwei Zhong, Report to Xuanhuai Sheng, November 1890, Chen, et al., *Hanyeping Gongsi*, 1: 21–22.
61 Chen et al., *Hanyeping Gongsi*, 1:22, 1:379, 383, 1:460–62, especially 1:333–34.
62 Project report, in Chen et al., *Hanyeping Gongsi*, 1:26, 1:48–49; Foreign engineer reports, in Chen et al., *Hanyeping Gongsi*, 1:27, 1:595.

63 For the impeachment reports, see Chen et al., *Hanyeping Gongsi*, 1:17, and 1:23.
64 For more impeachment reports, see Xulu et al., *Hanyeping Gongsi*, 1:452, 1:468, 1:476, 1:478, 1:547–48, 1:593, 1:598, 1:606, and 1:615–16.
65 Chen et al., *Hanyeping Gongsi*, 1:23–26. A tael is equal to about 1.3 ounces of silver.
66 Correspondence between Li Hongzhang and Zhang, in Xulu et al., *Hanyeping Gongsi*, 1:28–31, 1:42–45, 1:56.
67 Chen et al., *Hanyeping Gongsi*, 1:70–72.
68 Chen et al., *Hanyeping Gongsi*, 1:72.
69 Chen et al., *Hanyeping Gongsi*, 1:57–64, 1:66–69.
70 The Belgian appealed to the consulate first, to request help in getting paid by the HYP. Sheng, Report to Zhang, in Chen et al., *Hanyeping Gongsi*, 1:19.
71 Gongying Zheng, Reports to Sheng, June 1896 to August 1897, in Chen et al., *Hanyeping Gongsi*, 1:581–619.
72 The resignation reports, in Chen et al., *Hanyeping Gongsi*, 1:326, 1:333–34, 1:363, 1:413, 1:488–89, 1:491.
73 Sheng Chunyi, Report to Sheng, July 31, 1897, in Xulu et al., *Hanyeping Gongsi*, 1:614.
74 Xulu et al., *Hanyeping Gongsi*, 2:55–61.
75 Xulu et al., *Hanyeping Gongsi*, 1:376–77; for reports on the theft of thirteen boxes of explosives, see 1:418 and 1:689.
76 Xulu et al., *Hanyeping Gongsi*, 2:204–5.
77 For the contract of the first foreign loan, see Xulu et al., *Hanyeping Gongsi*, 2:24, 2:41–42, and 2:96–99.
78 HYP II, 486–89; the iron-ore and coal sale contracts and transportation records in Chen et al., *Hanyeping Gongsi*, 2:486–89.
79 For an introduction to Hubei science research and technology, see "Science," *HBSZ*, 17–59 (Wuhan: Hubei Renmin, 1990).
80 HBA, the monitoring reports on Gezhouba, SZ139–6–0358–1/2/3/6.
81 "Education section," *HBSZ*, 2–4, 207–32; "Education section," *WHSZ*, 3–4, 351–53.
82 Shaoguang Wang, *Failure of Charisma: The Cultural Revolution in Wuhan* (Oxford: Oxford University Press, 1995); "Administration," *HBSZ*, 411–13.
83 HPA, May 7, 1973, SZ139–06–480–004; December 5, 1972, SZ139–6–358–015; November 15, 1975, SZ139–6–628–001; January 12, 1986, SZ122–5–579–001.
84 Two intellectual figures, Huang Wangli and Ma Yinchu, are most credited in this respect. Shapiro, *Mao's War*.
85 "Science section," *HBSZ* (1990), 241–42.

Chapter 2: Groping for Stones to Cross the River

1 I am grateful to *China: An International Journal* and the East Asian Institute of the National University of Singapore for granting permission to reuse some material that appeared in Yun Liu, "Groping for Stones to Cross the River: Early Local Lessons from

Three Effluent Pollution Cases in Hubei in the 1960s–1980s," *China: An International Journal* 19, 3 (2021): 211–31.
2. Yiyi Lu, "Environmental Civil Society and Governance in China," *International Journal of Environmental Studies* 64, 1 (2007): 59–69. Usually, non-linear evolutions do not follow a predictable or sequential pattern, while featuring complex changes of multiple factors and producing unexpected outcomes.
3. Arthur P.J. Mol and Neil T. Carter, "China's Environmental Governance in Transition," *Environmental Politics* 15, 2 (2006): 149–70.
4. Effluent pollution received the most policy attention on the national scale during the 1990s. Xibing Huang et al., "Environmental Issues and Policy Priorities in China: A Content Analysis of Government Documents," *China: An International Journal* 8, 2 (2010): 229–30.
5. Jack A. Goldstone, "The Rise of the West – or Not? A Revision to Socio-Economic History," *Sociological Theory* 18, 2 (2000): 175–94. My writing is inspired by this revisionist approach, which rejects the Eurocentric view of history in which non-Western countries are Othered and portrayed as stagnant.
6. Internal reference circulation on investigation reports, compiled under the title of the HEP, Hubei Provincial Archives (HPA), SZ151–1–4; Issues 3 and 4, November 10–12, 1979.
7. Since 2015, the HPA has gradually declassified some investigation reports. Although they are unpublished, most of these files have been recatalogued and scanned, making them electronically accessible from the reading-room terminals at the archives, with staff assistance.
8. Memo on industrial waste pollution by iron and steel production, June 26, 1964, HPA, SZ115–02–0668–008.
9. Directive on public health warning for water contamination, January 10, 1968, HPA, SZ115–02–0868–001.
10. Warning on fishery production damaged by pollution, Ege ShuiDian [1972] 21, January 19, 1972, HPA, SZ113–6–89. "Ege" stands for "Hubei Provincial Revolution Committee." As cited here, all the local directives' original issuing catalogues would adhere to the following format: issuing institution [year] issue number. The HPA profiled these original issuing catalogues to build their HPA archival number; however, yet not all the archived documents have been designated with their in-house archive numbers in the HPA catalogue system, probably partly due to its growing workload.
11. HPA. Directive for strengthening industrial waste control, Ege [1975] 142, December 24, 1975.
12. Editorial comment, *Hubei Daily*, December 26, 2013.
13. The machinery company relocated to Jiangxia in the 1990s after years of intermittent protests by residents.
14. HPA. In 1980, there were a total of fifty-nine universities or colleges in Hubei, most of which were in Wuhan, registering about 600,000 students. Report on Dong Lake pollution, June 6, 1980, 6–7, SZ1–8–209–001. See also *Wuhan Municipal Yearbook* (Wuhan: Wuhan Publishing House, 1986).

15 HPA. Document collections on the water facilities issues in Dong Lake caused by pollution. Egeban [1976] 09, July 26, 1976; Egehuan [1975] 09, June 27, 1975; Egewen [1975] 070, March 27, 1975, directive, issued jointly by the Hubei Provincial Bureau of Health and SZ139–6–0699–013.
16 HPA. Report collection on urban pollution, Ege [1973] 73, May 7, 1973, HPA, SZ139–6–0480–004.
17 In December 2016, the Wugang Group merged with the Baosteel Group of Shanghai, creating the largest iron and steel syndicate in China today, the China Baowu Steel Group, with headquarters in Baoshan, Shanghai, and Qingshan, Wuhan.
18 HPA. Reports on water pollution, Ege [1973] 73, May 7, 1973, and Ege [1980] 61, SZ139–6–0480–004.
19 Directive for investigation update, Efa [1980] 61, July 18, 1980, HPA. "Efa" is short for Hubei [Provincial Government] issuance, and "E" in Efa is an abbreviation for Hubei, and "Efa" itself is short for Hubei [Provincial Government] issuance.
20 HPA. Reports on Hubei's environmental protection work, Efa [1980] 56, July 7, 1980, HPA, SZ1–8–209–005; *Changjiang Daily*, June 15, 1979; Ege [1973] 73, May 7, 1973.
21 Reports on Donghu pollution, Efa [1980] 61, July 18, 1980, HPA, SZ1–8–209–006.
22 HPA. Memo on the leader's talk to the provincial planning conference, June 26, 1980, HPA, SZ1–8–209–001; also refer to the report on water and land pollution, Ebanfa [1980] 27, June 5, 1980.
23 HPA. Report collection on Dong Lake pollution, May 7, 1973, HPA, SZ139–6–0480–004; Editorial, *Changjiang Daily*, June 15, 1979.
24 HPA. reports collection, SZ122–5–0579–001, Final assessment report of contamination of Dong Lake, Ekejianzi [1986] 001, January 3, 1986.
25 Public Facilities Section, *Wuhan Municipal Yearbook*; *Hubei Daily*, June 5, 2014, 6.
26 Section on Urban Public Water Facilities, *Wuhan Yearbook* (1986).
27 Reports on Donghu pollution; HPA, SZ1–9–606–004, Efa [1985] 29, November 12, 1985; Directive, Efa [1985] 19, July 9, 1985, HPA, SZ1–9–606–002/003; September 7, 1985, HPA, SZ1–9–607–007.
28 Memo on infrastructure update, March 14, 1985, HPA, SZ1–9–631–004. In 1985, most of Hubei's sixty-five counties still relied on their self-maintained wells.
29 HPA. Pollution report collection, SZ1–9–606–002; Efa [1985] 19, August 7, 1985.
30 HPA. Report of water facility construction for the universities by the Dong Lake area, December 12, 1985, HPA, SZ1–9–606–003; Ekejiaoban [1985] 065, December 10, 1985.
31 HPA. Collection of investigation reports on Ya'er Lake pollution; SZ139–6–0628–001.
32 Directive on Daye pollution, Egehuan [1979] 45, October 5, 1979; Pollution report on local lakes, Kewuyan [1979] 30, June 12, 1979.
33 Memo of people's letters, circulated in a *Situation Digest* by the *Renmin Ribao*, October 26, 1985, HPA, SZ1–9–880–002. The memo was first sent to the provincial branches and offices at Wuhan and then to lower levels.
34 Memo note, January 22, 1986, HPA, SZ1–9–880–003. The note was originally filed in December 1985.

35 HPA. Directive, Ege [1975] 119, November 15, 1975; Ege-jiji [1976] 456, October 15, 1976.
36 HPA. Budget note, Egejijizi [1976] 456, October 15, 1976; or Jijizhi [1976] 211, August 17, 1976.
37 HPA. Budget note, January–December 1979, SZ151-1-8; or Ewen [1978] 17, January 14, 1978.
38 HPA. Budget note, Eji [1982] 257, September 25, 1982.
39 HPA. Update on the Ya'er Lake case, Efa [1980] 61, July 18, 1980, SZ1-8-209-006.
40 HPA. Report collection on waste-processing fees; January 13, 1983, HPA, SZ93-03-0469-001; Also refer Report on approving waste-emission fees, Eyeneng [1983] 20, February 25, 1983.
41 HPA. Report on the Ya'er Lake pollution-control project in early stages, Egehuan [1979] 33, July 31, 1979; Egehuan [1978] 01, January 6, 1978.
42 HPA. Report on Ya'er Lake industrial pollution, December 19, 1975, SZ139-6-0628-002.
43 HPA. Report collection on Ya'er Lake pollution; SZ1-8-209-006.
44 *Yearbooks of the Hanyeping Factory* (Wuhan: Hubei Renmin Publishing House, 1986), HPA.
45 HPA. Reports on Gedian pollution, March SZ151-1-8; also see Egehuan [1979] 57, October 30, 1979; Egehuan [1979] 27, June 30, 1979.
46 The chemical and pharmaceutical industries have still dominated the local economy after the early 1990s.
47 HPA. Pollution report, Egehua [1979] 33, July 31, 1979, SZ151-1-7. Also see Egehua [1979] 20, May 4, 1979.
48 *Yingcheng City Yearbook* (1986–92), 84–89. Xiaogan: Yingcheng Yearbook Editorial Committee, 1995.
49 See the relay report of the internal reference memo, Egehuan [1979] 22, May 18, 1979, HPA.
50 Memo on a provincial environmental protection conference, Egehuan [1979] 22, May 12, 1979, 3–5; Overview investigation report on environmental pollution in Wuhan, Hubei, December 1978, HPA. The report was marked "Highly Secret."
51 HPA. Report on water pollution by hemp-fabric production by Hong Lake, Jinzhou, December 25, 1985, SZ1-9-880-001. A few reports also show some conflicts within the rural sector.
52 HPA. Directive of the National Bureau of Environmental Protection warnings to recurring water-pollution, Guohuan [1979] 17, May 12, 1979.
53 A negative correlation between national annual growth figures of industry and agriculture proves the latter's weak position. Renan Zhuang and Won W. Koo, "The Role of the Agricultural and Industrial Sectors in China's Economic Growth: Are They Twin Brothers?" *China: An International Journal* 6, 2 (2008): 299–300.
54 HPA. Directive note on food safety, Guofa [1974] 92, August 27, 1974.
55 HPA. Internal note on the Hubei environmental protection conference, November 12, 1979; Memo on Chen's speech in Wuhan, July 7, 1980, both profiled in SZ1-8-209-005.

56 HPA. Report on the Fu River report, Egehuan [1979] 48, October 10, 1979.
57 For reports on the Yichen arsenal pollution, see Egelongban [1973] 31, December 20, 1973; Ejungongji [1973] 565, December 19, 1973; and Egegongjiao [1973] 31, December 25, 1973, all in HPA, SZ26-04-046-002.
58 Report on follow-up investigations into pollution at Factory 525, by the provincial military industrial office, Egeguozhi [1975] 219, October 20, 1975.
59 Update report, Egehuan [1979] 20, May 4, 1979; also March 4, 1983, HPA, SZ93-03-0469-006.
60 Conference memo on Fu River pollution, Egehuan [1979] 24, June 15, 1979.
61 Gao Qi, "Public Interest Litigation in China: Panacea or Placebo for Environmental Protection?" *China: An International Journal* 16, 4 (2018): 47–75. See also Jian Lu and Chris King-Chi Chan, "Collective Identity, Framing and Mobilisation of Environmental Protests in Urban China: A Case Study of Qidong's Protest," *China: An International Journal* 14, 2 (2016): 102–22.
62 Michael Faure and Jing Liu address the complexity of compensating pollution victims. Michael G. Faure and Jing Liu, "Compensation for Environmental Damage in China: Theory and Practice," *Pace Environmental Law Review* 31, 1 (2014): 240–321.
63 Regarding these economic conflicts across local sectors, see Mark W. Skinner, Alun E. Joseph, and Richard G. Kuhn, "Social and Environmental Regulation in Rural China: Bringing the Changing Role of Local Government into Focus," *Geoforum* 34, 2 (2003): 267–81.
64 Internal memo on industrial waste control, Ege [1970] 57, April 15, 1970, 1–2.
65 Memo on the Hubei environmental protection conference, October 9, 1973, 19–20, HPA, SZH-422.
66 Guihuan Liu et al., "Eco-Compensation Policies and Mechanisms in China," *Review of European Community and International Environmental Law* 17, 2 (2008): 234–42.
67 Srini Sitaraman, "Regulating the Belching Dragon: Rule of Law, Politics of Enforcement, and Pollution Prevention in Post-Mao Industrial China," *Colorado Journal of International Environmental Law and Policy* 18, 2 (2007): 267–336; Srini Sitaraman, "Regulating the Environment: Assessing China's Domestic Environmental Law and Participation in International Treaties," *China Review* 6, 1 (2006): 183–96.
68 Unfortunately, as shown by the Hubei provincial court records for 1992–2005, these efforts at enforcement remained inadequate in the province. Xuehua Zhang, Leonard Ortolano, and Zhongmei Lu, "Agency Empowerment through the Administrative Litigation Law: Court Enforcement of Pollution Levies in Hubei Province," *China Quarterly* 202 (2010): 307–26. More field studies confirm the ineffectiveness of the centralized enforcement approach. Xuehua Zhang, "Implementation of Pollution Control Targets in China: Has a Centralized Enforcement Approach Worked?" *China Quarterly* 231 (2017): 749–74.
69 Rachel E. Stern, "On the Frontlines: Making Decisions in Chinese Civil Environmental Lawsuits," *Law and Policy* 32, 1 (2010): 79–103.
70 Rachel E. Stern, "From Dispute to Decision: Suing Polluters in China," *China Quarterly* 206 (2011): 294–312.

Chapter 3: Air Pollution and Soil Contamination

1. Arthur P.J. Mol and Neil T. Carter, "China's Environmental Governance in Transition," *Environmental Politics* 15, 2 (2006): 149–70; see also Lingxuan Liu, Bing Zhang, and Jun Bi, "Reforming China's Multi-Level Environmental Governance: Lessons from the 11th Five-Year Plan," *Environmental Science and Policy* 21 (2012): 106–11; and Olivia Bina. "Environmental Governance in China: Weakness and Potential from an Environmental Policy Integration Perspective." *China Review* 10, 1 (2010): 207–39. Portions of the present chapter are based on Yun Liu, "Voices of Protest against Industrial Pollution in Hubei, China, during the 1970s and 1980s," *Environment and History* 28, 4 (2022): 601–20.
2. Such a locally oriented perspective is also advocated in Maohong Bao, "Environmental History in China," *Environment and History* 10, 4 (2004): 475–99.
3. *Wuhan Shizhi (WHSZ)*, Wuhan: Wuhan Publishing House, 1986.
4. William T. Rowe, *Hankow: Conflict and Community in a Chinese City, 1796–1895* (Stanford: Stanford University Press, 1989).
5. This state-controlled retail network was connected by traditional commercial towns like Hankou and lasted until 1978; Hankou's commercial legacies survived out of this planning economy to some extent. Trade, *Hubei Shengzhi (HBSZ)* (Wuhan: Hubei Renmin, 1990), 62–75.
6. Transportation, *HBSZ* (1990), 102–16, 341–44.
7. *WHSZ* (1949–85)-Yange (Wuhan: Wuhan Publishing House, 1986), 110–11, 113–15.
8. An internal report on pollution in six major industrial cities in Hubei was initially classified as "Secret." Efa [1980] 56, July 7, 1980, Hubei Provincial Archives (HPA), SZ1–8–209–005.
9. HPA. Efa [1980] 1, the directive; 1 provincial government announcement, January 1, 1980, HPA.
10. HPA. Transcript of a speech by Pixian Chen, December 12, 1979, 21–29, SZH-422. Chen retired as secretary of the Hubei Provincial Committee of the CCP before taking the chair of the Provincial People's Consultative Congress.
11. HPA. Ege [1970] 57, April 15, 1970, SZ139–6–0628.
12. Memo on urban air pollution by American scientists, January 25, 1980, HPA, SZ151–1–2.
13. HPA. Guofa [1973] 158, November 13, 1973, also Ege [1974] 3, January 7, 1974.
14. HPA. Reference report on overseas environmental education, March 15, 1978, SZ118–4–678.
15. See Table 11–15: Annual statistics of pollution-control investment in Wuhan, in Wuhan Municipal Archives (WMA), ed., *Wuhan Yearbook: Wuhan Dadian* (Wuhan: Wuhan Publishing House, 1999, 2:1273.
16. WMA. Wuge [1979] 57, see also Wugehuan [1979] 23, citing the internal periodical *Wuhan Huanjing Baohu*, August 29, 1979, 5. Wugehuan stands for Wuhan Bureau of Environmental Protection.
17. HPA; Egehuan [1979] 27, June 30, 1979. Egehuan stands for the HEP as an issuer of reports/directives. See also HPA, SZH-422, 118–21.

18 An update on twelve urban air-pollution cases at Wuhan, *Wuhan Huanbao Jiaobao,* July 28, 1979, WMA, uncatalogued. The Wugang Group was the first of the twelve, followed by the Wuhan Heavy Machinery Group, which was also managed directly by the central agencies.
19 HPA; Ege [1977] 39, June 14, 1977, also Guofa [1973] 158, November 13, 1973.
20 Replies to investment requests, July 31, 1978, HPA, SZ43–05–1242–00.
21 *Wuhan Dadian* (*Wuhan Encyclopedia,* 1976–1998), volume 2, edited by the Wuhan Municipal Archive, Wuhan: Wuhan Publishing House, 1999, Table 11–15: Statistics of annual pollution-control project investment of industrial enterprises in Wuhan, 1273.
22 Refer the above citation in *Wuhan Dadian* (*Wuhan Encyclopedia,* 1976–1998). There is evidently a lagging pattern in the panel data as they show the total annual numbers of technological upgrading projects initiated within the reporting period. However, readers would probably not expect a high level of accuracy in these sectoral data.
23 Inquiry reports collection, recompiled in October 1984, 35–36, HPA, SZH-422.
24 HPA; Additional reports on local pollution petition cases, Egehuan [1979] 22, May 18, 1979, 3.
25 Local lake pollution reports, March 16, 1976, HPA, SZ128–1–579; Local lake pollution reports, June 7, 1978, HPA, SZ128–1–539.
26 The Zhongnan University of Law and Economics was disbanded between 1969 and 1971, while its main physical assets were assigned to the *Hubei Ribao,* the largest state-run daily newspaper in Hubei; it was reinstated in 1978 after nearly seven years of being suspended. In 1984, it spun off the Zhongnan College of Law and the Zhongnan University [College] of Finance and Economics. In 2000, the two schools were re-merged under the old name of the Zhongnan University of Law and Economics.
27 The cadre training institute, a predecessor of the Zhongnan University of Law and Economics, was shortly renamed "Hubei University" in 1978. Its main campus, in a crowded area of Wuchang, was temporarily designated to the *Hubei Ribao* during the Cultural Revolution and was restored in the mid-1980s.
28 Report on the letter of appeal, September 6, 1986, HPA, SZ1–9–880–004.
29 Follow-up inquiry report, August 4, 1986, 34–36, HPA, SZ1–9–880–003. The report notes that the investigation copy was printed, relayed, and referred to the university's CCP commission.
30 Philip C.C. Huang, "'Public Sphere'/'Civil Society' in China? The Third Realm between State and Society," *Modern China* 192, 2 (1993): 216–40.
31 Memo for the early incidents in Qingshang, Egehuan [1979] 33, July 31, 1979, HPA.
32 Efa [1980] 61, July 18, 1980, HPA, SZ1–8–209–006. These follow-up reports also present no specific pollution details but evidence of inadequate investment.
33 Egehuan [1979] 33, July 31, 1979, 3–6, HPA.
34 Memo of key points planning for environmental protection, December 15, 1974, 3–8, HPA, SZH-422.
35 Guohuan [1979] 30, April 20, 1979, HPA; Egehua [1979] 22, February 12, 1979, 7.
36 *Hubei Nianjian* (Wuhan: Hubei Renmin, 1995), 317–18, HPA; *Huangshi Shizhi* (*Huangshi Gazetteer*) (Beijing: Zhonghua Shuju, 2001), 1:99–100.

37 Report on utilizing lake resources in Hubei, Efa [1979] 131, November 29, 1979, HPA.
38 Transcript of speech by Vice-Governor Tian Ying, January 23, 1980, 61–69, HPA, SZH-422.
39 Internal memo for Han Ningfu, secretary of the Hubei Provincial Committee of the CCP, January 28, 1980, HPA, SZH-422. See also Ezhenfa [1981] 33, March 13, 1981. Another memo readdressed the same facts by the former secretary of the Hubei Provincial Committee of the CCP, Pixian Chen. Efa [1980] 1, December 19, 1979.
40 Simon Avenell, "From Fearsome Pollution to Fukushima: Environmental Activism and the Nuclear Blind Spot in Contemporary Japan," *Environmental History* 17, 2 (2012): 244–76.
41 Ezhenfa [1981] 33, March 13, 1981, HPA; Efa [1980] 1, December 19, 1979.
42 Memo of provincial environmental conference, Egehuan [1979] 22, May 12, 1979, HPA.
43 Supplementary memo on the speech on January 28, 1980, 71–78, HPA, SZH-422.
44 HPA; Ewen [1978] 17, March 10, 1978, a reply report to the early lake pollution reports filed on January 24, 1978. As one of the most notable figures of those who received the reply report, Li Xiannian was president of the People's Republic of China from 1983 to 1988 and was then chairman of the Chinese People's Political Consultative Conference until 1992.
45 Transcript of internal talk by Vice-Governor Tian Ying, April 11, 1982, 82–91, HPA, SZH-422.
46 Local notables section, *Dawu County Gazetteer* (Dawu: Dawu Yearbook Editorial Office, 1996), 30–40, 630, 740–75.
47 HPA. Ezhenwen [1981] 51, May 9, 1981; see also *Dawu County Gazetteer* (1996), 285–87.
48 Attached to a memo for a HEP internal conference, Egehuan [1979] 22, May 12, 1979, HPA.
49 *Dawu County Gazetteer* (1996), 285–87.
50 For a case involving letters of appeal by "anonymous" local peasants, see the HEP internal circulation, issue 2, February 6, 1980, and issue 4, April 12, 1980, HPA, SZ151-1-21.
51 Jonathan D. Spence, *The Search for Modern China*. 1st ed. New York: Norton, 1990. Public debates on the exam-oriented educational system of the PRC have persisted into the twenty-first century, since the return of the exam-based selection policy for entry into the civil service.
52 "Zha and the Restoration Stories of College-Entry Exams," *People's Daily*, overseas edition, October 10, 2014, section 10; Obituary for Zha, *Changjian Daily*, August 1, 2019.
53 The same report by the *People's Daily*, overseas edition, October 10, 2014, section 10.
54 Daoyu Liu, *Confession of a University President* (Wuhan: Changjian Press, 2005).
55 "Liu Daoyu Is Designated to Be the President of Wuhan University," *People's Daily*, August 22, 1981, front page.
56 *Guangming Daily*, December 10, 2008. Since 1949, the readers of this national newspaper have been teachers and intellectuals.

57 On some recent unofficial occasions, Tang has denied his leadership role in advocating "education industrialization," and this once-popular term, with a whole package of state-endorsed policies, has incurred many harsh criticisms regarding its overly commercialization or capitalization tendency.
58 Statistics Bureau of Wuhan, *Wuhan Statistical Yearbook of 2015* (Wuhan: Wuhan Publishing House, 2015), 407–9.
59 Bryan Tilt, *The Struggle for Sustainability in Rural China: Environmental Values and Civil Society* (New York: Columbia University Press, 2010).
60 For a study measuring the "receptivity" of authorities to non-official views and arguing such effects would depend most on the perceived quality of state–society relations, see Tianguang Meng, Jennifer Pan, and Ping Yang. "Conditional Receptivity to Citizen Participation: Evidence From a Survey Experiment in China." *Comparative Political Studies* 50, 4 (2017): 399–433.
61 Kenneth Pomeranz, "The Transformation of China's Environment, 1500–2000," in *The Environment and World History*, ed. Edmund Burke and Kenneth Pomeranz (Berkeley: University of California Press, 2009), 118–64.
62 Helen Dunstan, "Official Thinking on Environmental Issues and the State's Environmental Roles in Eighteenth-Century China," in *Sediments of Time: Environment and Society in Chinese History*, ed. Mark Elvin and and Curirong Liu (Cambridge: Cambridge University Press, 1998), 585–614. See also Pomeranz 2007.
63 Xuehua Zhang, "China's Environmental Administrative Enforcement System," *Environmental Law Reporter: News and Analysis* 41, 10 (2011): 10890–94.
64 Michael G. Faure and Jing Liu, "Compensation for Environmental Damage in China: Theory and Practice," *Pace Environmental Law Review* 31, 1 (2014): 240–321; see also Guihuan Liu et al., "Eco-Compensation Policies and Mechanisms in China," *Review of European Community and International Environmental Law* 17, 2 (2008): 234–42.
65 Peter Ho and Richard Louis Edmonds, "Perspectives of Time and Change: Rethinking Embedded Environmental Activism in China," 2007.

Chapter 4: Struggles for Policy Implementation

1 A significant portion of this chapter was originally published in Yun Liu, "Struggling for Policy Implementation: The Establishment of Provincial Environmental Agencies in the 1970s and 1980s in Hubei, China," *Twentieth-Century China* 44, 1 (2019): 98–115.
2 For discussions of such continuity and rupture, see Helen Dunstan, "Official Thinking on Environmental Issues and the State's Environmental Roles in Eighteenth-Century China," in *Sediments of Time: Environment and Society in Chinese History*, ed. Mark Elvin and Curirong Liu (Cambridge: Cambridge University Press, 1998), 585–614.
3 Elizabeth Economy, "Environmental Governance: The Emerging Economic Dimension," *Environmental Politics* 15, 2 (2006): 171–89; Elizabeth Economy, *The River Runs Black: The Environmental Challenge to China's Future* (Ithaca: Cornell University Press, 2010).

4 Richard Louis Edmonds, "The Evolution of Environmental Policy in the People's Republic of China," *Journal of Current Chinese Affairs* 40, 3 (2011): 13–35.
5 The term "mass movement" in Chinese can refer either to state-sponsored campaigns to mobilize "the masses" for certain state-defined goals or to social movements generated from below. This chapter uses "state campaign" for the former.
6 Because China is such a vast, ancient, and culturally diverse country riddled with what may be considered complex paradoxes to outsiders, the priorities of subnational policies, and their implementation, would commonly vary from region to region. Abigail Jahiel, "The Organization of Environmental Protection in China," *China Quarterly* 156 (1998): 759; see also Robert B. Marks, "Why China?" *Environmental History* 10, 1 (2005): 56–58.
7 David Allen Pietz, *The Yellow River: The Problem of Water in Modern China* (Cambridge, MA: Harvard University Press, 2015), particularly Chapter 4, "Making the Water Run Clear," 130–93, and Chapter 5, "Creating a Garden in North China Plain," 194–257, regarding the role of the Yellow River Water Resource Commission.
8 Sebastian Heilmann, "From Local Experiments to National Policy: The Origins of China's Distinctive Policy Process," *China Journal* 59 (2008): 2–3.
9 As revealed by interviews in Sigrid Schmalzer, *Red Revolution, Green Revolution: Scientific Farming in Socialist China* (Chicago: University of Chicago Press, 2016).
10 Many scholars stress regional remedial policies. See, for example, Eduard B. Vermeer, "Industrial Pollution in China and Remedial Policies," *China Quarterly* 156 (1998): 952–85.
11 Jahiel, "The Organization of Environmental Protection," 760–65.
12 The Hubei Provincial Rural Energy Office operated under many aliases – such as "Hubei Office of Rural Energy," "Provincial Office of Rural Energy," "Provincial Leadership Office of Biogas Utilization," "the Biogas Office," "the Provincial Office of Rural Energy" or "Hubei Provincial Office of Rural Energy" – in various and/or overlapping stages.
13 Vaclav Smil, *The Bad Earth: Environmental Degradation in China* (Armonk, NY/ London: M.E. Sharpe/Zed Press, 1984).
14 Judith Shapiro, *Mao's War against Nature: Politics and the Environment in Revolutionary China* (Cambridge, MA: Cambridge University Press, 2001); Judith Shapiro, *China's Environmental Challenges* (Malden, MA: Polity Press, 2012).
15 Vermeer, "Industrial Pollution," 984–85; see also Richard Louis Edmonds, "Studies on China's Environment," *China Quarterly* 156 (1998): 725–32.
16 HPA. Ege [1974] 57, November 30, 1974.
17 HPA. Guofa [1979] 237, October 8, 1979. The most recent revision of the national law of Environmental Protection was passed in April 2014.
18 HPA. June 1983, SZH-427.
19 HPA. Ege [75] 070, June 23, 1975, SZ99–06–0205–002.
20 Ege [1970] 57, April 15, 1970.
21 Ege Shuidian [1972] 21, January 19, 1972, HPA, SZ113–6–89.

22 Supplementary investigation report by Wuhan DPMS, EgeShuidian [1971] 74, December 21, 1971; Conference memo, HEP, Egehuan [1979] 22, May 21, 1979.
23 HPA. Directive report collections; June 26, 1964, SZ115–02–0668–008.
24 January 10, 1968, HPA, SZ115–02–0868–001.
25 HPA. Reply to the early reports, Ege [1975] 070, July 23, 1975.
26 HPA. Ege [1973] 73, May 7, 1973; Report on Dong Lake pollution, May 3 1972. For a more recent report addressing this case, see *Changjiang Daily,* June 15, 1979.
27 HPA. Ege [1974] 3, January 7, 1974, SZ99–06–0205.
28 HPA. August 29, 1973, SZ99–06–0205–001.
29 HPA. Report to the central government, with a record of 10,000 attendees, Ege [1973] 191, November 16, 1973. SZ115–05–0032–004.
30 Provincial conference of environmental protection, October 9, 1973. SZ115–05–0032–004.
31 HPA. Ege [1974] 57, November 30, 1974.
32 HPA. Update report, December 7, 1974, SZ99–06–0205–007.
33 HPA. Egewei [1976] 027, October 6, 1976, SZ115–05–0071–0021.
34 HPA. Ewen [1977] 96, July 7, 1977.
35 HPA. Ege [1979] 95, June 23, 1979, 116.
36 HPA. Egeban [1996] 18, November 31, 1996; see also Ebian [2006] 83, December 14, 2006.
37 Smil, *The Bad Earth,* 152–53.
38 Smil, *The Bad Earth,* 198.
39 E. Ariane Van Buren, "Biogas beyond China: First International Training Program for Developing Countries," *Ambio* 9, 1 (1980): 14.
40 HPA. Report of the national biogas conference, June 19, 1975, SZ139–6–603–3.
41 Heilmann, "From Local Experiments to National Policy," Figure 1: Establishing "Model Experiments": A comparison of Maoist and Dengist approaches from local experiments to national policy.
42 T.K. Moulik, "The Biogas Program in India and China," *Ambio* 14, 4–5 (1985): 288–90.
43 Moulik, "The Biogas Program," 291.
44 This national investment statistics for China's biogas economy is from Zuzhang Xia. *Domestic Biogas in a Changing China: Can Biogas Still Meet the Energy Needs of China's Rural Households?* (London: International Institute for Environmental and Development, 2013).
45 Yu Chen et al., "Household Biogas Use in Rural China: A Study of Opportunities and Constraints," *Renewable and Sustainable Energy Reviews* 14, 1 (2010): 545–49. The authors provide historical data from 1973 to 2006, which show that the national figure for digesters has increased since 1984, gradually at first but then exponentially from the late 1990s.
46 Smil, *The Bad Earth,* 199.
47 Kenneth Pomeranz, "Chinese Politics and Environmental History," *Environmental History* 12, 2 (2007): 352–54; Kenneth Pomeranz, *The Great Divergence: China, Europe,*

and the Making of the Modern World Economy (Princeton and Oxford: Princeton University Press, 2009). Also refer to E. Elena Songster, *Panda Nation: The Construction and Conservation of China's Modern Icon* (Oxford: Oxford University Press, 2018), Chapter 2, National Building and the Nature of Communist Conservation.

48 Hubei Provincial Bureau of Industries, report by the research team of bio-energy utilization, December 3, 1957, HPA, SZ90–2–976–1.
49 HPA. Status report, June 1957, Page 12, SZ90–2–976–1; also refer to the Japanese biogas-device pamphlet, SZ90–2–976–4.
50 HPA. Follow-up report, December 1957, 35–46, SZ90–2–96–2/3.
51 HPA. Letter from Biwu Dong, issued on April 17, 1960, filed on April 27, SZ122–1–0021–2.
52 HPA. Internal memo, March 20, 1960, SZ122–1–0021–1. The March 15, 1960, issue of the *Hubei Daily* published an article on how much labour and materials would be needed to build a biogas generator.
53 HPA. Notice for field-conference of biogas and microwave technology, SZ122–0021–6.
54 HPA. Report on the staged biogas expansion of Hubei, SZ122–2–231.
55 HPA. Report on the Shayang Farm, January 8, 1968, collected in SZ122–2–364.
56 HPA. Administrative reply, Ege [1979] 130, September 7, 1979.
57 HPA. Administrative reply, Ezhenban [1983] 89, November 19, 1983.
58 HPA. Status report. July 12, 1975, SZ139–6–603–3.
59 HPA. Directive report. Egezhao [1979] 21, October 13, 1979.
60 *Hubei EPMS and Environmental Protection Institute Yearbook* (June 1983), 3–4.
61 Report, the National Department of Agriculture, Nongyebu [79] Nongke 61.
62 There is a nuanced difference in terms of bureaucratic rank between a bureau and a department (or a ministry). This small difference, within the same provincial government body, can imply half a level of ranking gap, and the bureau may have a relatively lower status, but not necessarily in cases. If a bureau is operated within or affiliated with a ministry or department, the rank of its head is usually equivalent to that of a vice-director of a ministry or department. However, a vice-director with seniority can hold the same rank of a chief director. In more independently operated bureaus, the head can have the same rank as the head of a department or ministry.
63 Refer the reports published in 2015, the Yearbook of Hubei Agriculture, the Hubei Provincial Department of Agriculture (Wuhan: Hubei Renmin Publishing House, 2015).
64 Jahiel, "The Organization of Environmental Protection," Figure 1: The Chinese Environmental Protection Apparatus (March 1998); it illustrates a national roadmap of environmental protection and industrial waste pollution regulations, also refers to Lester Ross, "China: Environmental Protection, Domestic Policy Trends, Patterns of Participation in Regimes and Compliance with International Norms." *China Quarterly* 156 (1998): 811–12.
65 Dieter Grunow, "Structures and Logic of Environment Protection Implementation and Administration in China," *Journal of Current Chinese Affairs* 40, 3 (2011): 42, 70–72.

66 For instances of continuing investment, see HPA, Egehuan [1979] 28, co-issued with Erjijiang [1979] 183, July 9, 1979. See also Ege-Jiji [1976] 093, March 9 1976.
67 HPA. HEP policy proposal attachment, Egehuan [1979] 22, May 18, 1979.
68 HPA. Guohuan [1978] 20, October, 17, 1978; Guohuan [1977] 3, April 14, 1977.
69 HPA. Funding request report for Sanfei control, March 9, 1979, SZ69–7–602–001. Another early report to request funding is EgeJicheng [1972] 233, SZ99–06–0098–05.
70 HPA. Provincial administrative approval to the Sanfei report, Ege [1972] 125, SZ99–06–0098–002/003.
71 HPA. Semi-annual work schedule report, Egehuan [1979] 33, June 31, 1979.
72 HPA. Report on pollution-control reimbursement. SZ93–03–0469–003. The rewards included reimbursing emission fees and appropriations.
73 HPA. Directive report, Guohuan [1978] 20, SZ93–03–0469–005; see also Update report, Egehuan [1979] 27.
74 HPA. Report by Hubei to fund Sanfei control projects, Egeji [1972] and Egejicheng [1972], SZ99–06–0098–006; see also Ege [1977] 39, April 14, 1977, SZ139–6–0740–005.
75 Egehuan [1979] 65, November 27, 1979. See also Report for the national conference on environmental protection, Guohuanban [1979] 30, April 30, 1979, 8–9.
76 Xuehua Zhang, Leonard Ortolano, and Zhongmei Lu, "Agency Empowerment through the Administrative Litigation Law: Court Enforcement of Pollution Levies in Hubei Province," *China Quarterly* 202 (2010): 307–26.

Chapter 5: The Right to Pollute

1 Senior party officials in Hubei stressed this inter-sectoral conflict. One example appears in a national policy report delivered at a central-level environmental protection conference on October 9, 1973, by Jun Yan, vice-director of the Hubei Provincial Revolutionary Committee. His status at the time was equivalent to that of deputy premier of Hubei. 1983, 21, Hubei Provincial Archives (HPA), SZH-422.
2 Many economists distinguish ecological economics from resource and environmental economics. See, for example, Mick Common, "Economics and the Natural Environment: A Review Article," *Journal of Economic Studies* 25, 1 (1998): 57–73; and Clive L. Spash and Anthony Ryan, "Economic Schools of Thought on the Environment: Investigating Unity and Division," *Cambridge Journal of Economics* 36, 5 (2012): 1091–121.
3 Juan Martinez-Alier, *Ecological Economics: Economics, Environment, and Society* (Oxford: Basil Blackwell, 1987).
4 Steve Mark Cohn, *Competing Economic Paradigms in China: The Co-evolution of Economic Events, Economic Theory and Economics Education, 1976–2016*. Oxfordshire: Routledge, 2017.

5 Fugu Cheng. "Shengtai Jingjixue Yuanliu" [Origins of Ecological Economics], *Economic Research Journal* 9 (1983): 43–49.
6 Yingguang Jiang, "Youguan Shengtai Jingji De Jige Lilun Wenti" [Several Theoretical Questions in Ecological Economics], *Economic Research Journal* 10 (1983): 56–60. See also Dixin Xu, "Shixian Sihua Yu Shengtai Jingjixue" [Four Modernizations and Ecological Economics], *Economic Research Journal* 11 (1980): 14–18. Chuandong Ma, "Tigao Nengyuankaifa Liyong De Jingji Shengtaixiaoyi: [Promoting Ecological-Economic Benefit from Upgrading Energy Utilization], *Economic Research Journal* 2 (1985): 58–62. Zhonglian Sun, "Kuangchanziyuan De Youchang Shiyong" [Compensation Costs in Using Mineral Resources], *Economic Research Journal* 2 (1986): 70–75.
7 Tian Shi, "Ecological Economics in China: Origins, Dilemmas and Prospects," *Ecological Economics* 41, 1 (2002): 5–20.
8 Also refer to Dongshen Cheng, "Huanjing Jingjixue Chutan" [Origins of Environmental Economics], *Social Sciences in Yunnan* 6 (1982): 38–47. See also Yushi Mao, "Huanjing Jingjixue Zhong Sange Jiben Wenti" [Three Fundamental Questions in Environmental Economics], *Economic Research Journal* 7 (1982): 60–63.
9 Zhufu Fu, "Youguan Zhongguo Jingjishi De Ruogan Teshu Wenti" [Three Fundamental Questions in Environmental Economics], *Economic Research Journal* 7 (1978): 49–61. Zhongping Yang. "Yao Jiaqiang Dui Jingjishi De Yanjiu" [Strengthening the Researches on Economic History], *Economic Research Journal* 10 (1981): 19–20. Tongfong Qiao, "Jinnianlai Shengtai Jingji Wenti Taolun Zongshu" [an Overview on Ecological – Economic Problems in Recent Years], *Economic Research Journal* 2 (1982): 77–80. Junjian Jin, "Jiaqiang Zhongguo Jingjishi Yanjiu Shi Fazhan Jingji Xueke De Yixiang Zhongyao Zhanlüe Renwu" [A Strategic Task to Strengthen the Researches of Chinese Economic History for the Development Economics Studies], *Economic Research Journal* 10 (1983): 61–68. Shihua Liu, "Lun Shengtai Jingji Xuqiu" [Demand of Ecological Economy], *Economic Research Journal* 4 (1988): 77–79.
10 Xu held many other offices: he was minister of the state administration for industry and commerce of China, deputy head of the CCP United Front Work Department, and chairman of the CEES, which was founded in 1984.
11 See Chaozhun Zhang, "Guanyu Zhongguoshi Xiandaihua Daolu De Tantao" [Exploring the Chinese Modernization Path], *Economic Research Journal* 9 (1979): 3–9. Also Dixin Xu. "Youguan Nongye Jingji De Jige Wenti" [Several Problems in Agricultural Economics], *Economic Research Journal* 12 (1978): 8–16. Zhongyi Liu and Yaochuang Liu, "Shixian Nongye De Liangge Liangxing Xunhuan" [Materializing Two Benign Cycles of Agriculture], *Economic Research Journal* 5 (1982): 7–12.
12 HPA. Provincial report requesting mediation of imposing violation fines, Ehuanguan [1991] 80; November 19, 1991.
13 HPA. Ege [1978] 28, January 23, 1978, SZ139–6–0832–002.
14 HPA. Ege [1974] 03, January 7, 1974, SZ99–06–0205–001.
15 Provincial environmental protection construction subsidy investment report, July 31, 1978, HPA, SZ43–05–1242–001. See also Investment adjustment for Hubei's provincial environmental protection subsidy, October 10, 1978, HPA, SZ43–05–1242–005.

16 HPA. Notice of some probation codes in collecting emission or discharge fees, Ezhenfa [1982] 70, June 26, 1982. See also Ecai [1982] 67, December 28, 1982.
17 E.P. Thompson, "The Moral Economy of the English Crowd in the Eighteenth Century," *Past and Present* 50 (1971): 135.
18 Kjell Arne Brekke, Ragnhild Haugli Braaten, and Ole Rogeberg, "Buying the Right to Do Wrong – an Experimental Test of Moral Objections to Trading Emission Permits," *Resource and Energy Economics* 42 (2015): 110–24.
19 For examples of the many articles that provide an overview of China's one-child policy, see Martin King Whyte, Wang Feng, and Yong Cai, "Challenging Myths about China's One-Child Policy," *China Journal* 74 (2015): 144–59; Therese Hesketh, Xudong Zhou, and Yun Wang, "The End of the One-Child Policy: Lasting Implications for China," *Jama* 314, 24 (2015): 2619–20; and Yi Zeng and Therese Hesketh, "The Effects of China's Universal Two-Child Policy," *The Lancet* 388, 10054 (2016): 1930–38.
20 David de la Croix and Axel Gosseries, "The Natalist Bias of Pollution Control," *Journal of Environmental Economics and Management* 63, 2 (2012): 271–87.
21 Janos Kornai, "The Soft Budget Constraint," *Kyklos* 39, 1 (1986): 3–30.
22 Justin Yifu Lin and Guofu Tan, "Policy Burdens, Accountability, and the Soft Budget Constraint," *American Economic Review* 89, 2 (1999): 426–31.
23 Hubei Provincial Committee of the CCP, Internal proposal for an official memo of Hubei's significant events (March 1971– March 1980), particularly for the period of 1979–80, 65–69, HPA, SZ1-8-661-001; Report on the speech of the vice-premier of Hubei, Han Linfu (韩宁夫), in an internal meeting for the provincial long-term planning agenda, Ebanfa [1980] 71, June 26, 1980, HPA, SZ1-8-209-001. See also Internal report on Hubei economic plans in 1980–85, Ebanfa [1985] 13, March 12, 1985, HPA, SZ1-9-629-001.
24 Gang Li, "Kexuejishu Shi Guanxi Guojia Shengsicunwang De Yige Zhongyao Yinsu– Xuexi Zhongguo Jindai Jingjishi De Yidian Tihui" [Science Is the Important Factor for National Survival – Some Reflections on the Study of Chinese Modern Economic History], *Economic Research Journal* 5 (1978): 60–64.
25 Mao was reputed for his pioneering introduction of neo-classical economics. Yushi Mao, "Zeyou Fenpei Yuanli Jianjie" [A Short Introduction to the Principles of Optimal Allocation], *Economic Research Journal* 12 (1980): 65–68.
26 Mao (1982), "Sange Jiben Wenti" [Three Fundamental Questions in Environmental Economics], 60.
27 Ma (1985), "Nengyuankaifa Liyong" [Promoting Ecological-Economic Benefit from Upgrading Energy Utilization].
28 Ming Xian and Lu Bing, "Maoyushi De Jingjixue Rensheng" [Mao Yushi's Life in Economics], *China Reform* 11 (2013): 37–42.
29 For example, the views of Xu Dixin (1980), "Four Modernizations and ecological economics" are similar to those of Yushi Mao (1982).
30 Manhong Shen, "Lun Huanjing Jingji Shouduan" [Exploring the Market – Based Instruments for an Economic – Environmental Policy Framework], *Economic Research Journal* 10 (1997): 54–61.

31 Xiao Zhang, "Zhongguo Huanjing Zhengce De Zongti Pingjia" [An Overall Assessment on China's Environmental Policy], *Social Sciences in China* 3 (1999): 88–99. Also refer Keguo Li, "Zhongguo De Huanjing Jingji Zhengce" [Chinese Environmental Economy Policy], *Ecological Economy* 11 (2000): 39–42.
32 Michael Palmer, "Environmental Regulation in the People's Republic of China: The Face of Domestic Law," *China Quarterly* 156 (1998): 788–808; Lester Ross, "China: Environmental Protection, Domestic Policy Trends, Patterns of Participation in Regimes and Compliance with International Norms," *China Quarterly* 156 (1998): 809–35.
33 Tilt, Bryan. "Industrial Pollution and Environmental Health in Rural China: Risk, Uncertainty and Individualization." *The China Quarterly* 214 (2013): 283–301. Robert F. Ash and Richard Louis Edmonds, "China's Land Resources, Environment and Agricultural Production," *China Quarterly* 156 (1998): 836–79; Kate E. Swanson, Richard G. Kuhn, and Wei Xu, "Environmental Policy Implementation in Rural China: A Case Study of Yuhang, Zhejiang," *Environmental Management* 27, 4 (2001): 481–91.
34 Ali Douai, Andrew Mearman, and Ioana Negru, "Prospects for a Heterodox Economics of the Environment and Sustainability," *Cambridge Journal of Economics* 36, 5 (2012): 1019–32.
35 Report to the central government, Ege [1973] 191, November 16, 1973, 27–30, HPA, SZH-422.
36 Eduard B. Vermeer, "Industrial Pollution in China and Remedial Policies," *China Quarterly* 156 (1998): 985.
37 HPA. Relay Report, Guohuan 1981 [81] 9, SZ93–03–0222–004; see also HPA, Eyeneng [1981] 41, March 27, 1981; and SZ93–03–0222–005.
38 Smil also feels that effective environmental protection in China was obstructed by the "political implications of foreign technology acquisition," which was a huge cloud "hanging over every major decision in China." Vaclav Smil (1998). Given China's rapid pace in importing advanced Western technology from the time of this observation, it now seems overstated.
39 Xuehua Zhang, Leonard Ortolano, and Zhongmei Lu, "Agency Empowerment through the Administrative Litigation Law: Court Enforcement of Pollution Levies in Hubei Province," *China Quarterly* 202 (2010): 307–26.
40 Pierre Desrochers, "Did the Invisible Hand Need a Regulatory Glove to Develop a Green Thumb? Some Historical Perspective on Market Incentives, Win-Win Innovations and the Porter Hypothesis," *Environmental and Resource Economics* 41, 4 (2008): 519–39.
41 Michael E. Porter and Claas van der Linde, "Toward a New Conception of the Environment-Competitiveness Relationship," *Journal of Economic Perspectives* 9, 4 (1995): 97–118. See also Cynthia A. Montgomery and Michael E. Porter, eds., *Strategy: Seeking and Securing Competitive Advantage* (Boston: Harvard Business School Press, 1991).
42 Elizabeth Economy, "Environmental Governance: The Emerging Economic Dimension," *Environmental Politics* 15, 2 (2006): 171–89. See also Arthur P.J. Mol and Neil

T. Carter, "China's Environmental Governance in Transition," *Environmental Politics* 15, 2 (2006): 149–70; and Srini Sitaraman, "Regulating the Belching Dragon: Rule of Law, Politics of Enforcement, and Pollution Prevention in Post-Mao Industrial China," *Colorado Journal of International Environmental Law and Policy* 18, 2 (2007): 267–336. All these articles highlight China's constitutional or political constraints on environmental policy.

43 HPA. Emergency notice by the Provincial Revolution Committee, Ege [1970] 57, April 10, 1970.
44 HPA. Internal speech memos for Kong Qingde and Yan Jun, October 9, 1973, 9–10, SZH-422.
45 HPA. An attached note regarding the HEP budget coordination, from the Provincial Department of Public Finance to its central supervisor, April 8, 1975, SZ139–6–0653–003.
46 HPA. Budget report for the sewage system upgrading project of the East Lake at the Qingshan river port, Ege [1975] 007, January 12, 1975, SZ139–6–0628–005.
47 HPA. Egewen [1975] 070, July 23, 1975. SZ139–6–0653–003.
48 HPA. Ege [1977] 39, June 14, 1977; Egehuan [1975] 01, January 7, 1975.
49 HPA. Letter, April 11, 1980, included in internal memo of the HEP, SZ151–1–21.
50 Report by the Wuhan Branch of the Chinese Academy of Science, Wuke [1979] 30, June 12, 1979.
51 HPA. Guohuanzi [1978] 20, October 17, 1978. See also Egehuan [1978] 43, November 20, 1978. The National Construction Committee joined this group in this later series of directives.
52 HPA, SZ43–05–1242–001 and 003, July 31, 1978, these profiled records contain examples of investment request reports and official responses allocating annual investment for environmental protection.
53 Douglass C. North, *Understanding the Process of Economic Change*. Princeton Economic History of the Western World (Princeton: Princeton University Press, 2005).
54 Randolph M. Lyon, "Auctions and Alternative Procedures for Allocating Pollution Rights," *Land Economics* 58, 1 (1982): 16–32.
55 Charles W. Howe and Dwight R. Lee, "Priority Pollution Rights: Adapting Pollution Control to a Variable Environment," *Land Economics* 59, 2 (1983): 141–49.
56 Donald R. Ryan, "Transferable Discharge Permits and the Control of Stationary Source Air Pollution: Reply," *Land Economics* 59, 1 (1983): 126–27.
57 Peter Perdue, "Zhongguo Huanjingshi Yangjiu Xiangzhuang Ji Qushi" [Trends and status of Chinese environmental history research], *Jianghan Tribune* 5 (2014): 38–40.
58 Yu-Bong Lai, "The Optimal Distribution of Pollution Rights in the Presence of Political Distortions," *Environmental and Resource Economics* 36, 3 (2007): 367–88.
59 *Hubei Daily*, November 2, 2006; *Beijing News*, November 2, 2006.
60 Jethro Mullen, "Tons of Poisoned Fish Clog River in China's Hubei Province," *CNN*, September 5, 2013, http://www.cnn.com/2013/09/05/world/asia/china-river-dead-fish/.
61 Bo Zhang and Cong Cao, "Four Gaps in China's New Environmental Law: Implementation and Accountability Will Remain Challenging, Especially at the Local

Level, Warn Bo Zhang and Cong Cao," *Nature* 517, 7535 (2015): 433–35; Lei Zhang, Arthur P.J. Mol, and David A Sonnenfeld, "The Interpretation of Ecological Modernisation in China," *Environmental Politics* 16, 4 (2007): 659–68.

CHAPTER 6: UPDATING ENVIRONMENTAL GOVERNANCE IN WUHAN

1. Report for the Donghu Sewage Interception Project, Section of Rural-Urban Infrastructure and Environmental Protection, *Wuhan Nianjian (WHNJ)* (Wuhan: Wuhan Publishing House, 1987), 88.
2. Report about the Huangxiao River Environmental Re-modification Project, Section of Rural-Urban Infrastructure, *WHNJ* (1987), 88.
3. Report about the Huangxiao River Environmental Re-modification Project. The information in the next few paragraphs also comes from this report.
4. Many newspaper articles discuss the play. For an example, see Cultural Section, *Beijing Daily*, January 12, 2022.
5. Overview report on the year 1986 environmental work achievement, Section of Rural-Urban Infrastructure and Environmental Protection, *WHNJ* (1987), 101–2.
6. Report on the research project Waterbody Organic Pollution Leading to Mutagenicity, Section of Rural-Urban Infrastructure and Environmental Protection, *WHNJ* (1987), 103.
7. Review report on Environmental Prediction Research, Section of Rural-Urban Construction and Environmental Protection, *WHNJ* (1987), 103–4.
8. Reports on dust-control technological innovation by the Wugang Company and the Wuhan Cement Factory, *WHNJ* (1987), 104.
9. Review report on drinking water source protection, Section of Rural-Urban Construction and Environmental Protection, *WHNJ* (1987), 107.
10. Report on international environmental research exchange, *WHNJ-Enviromental Protection Section* (Wuhan: Wuhan Publishing House, 1988), 98. The research exchange, a three-year project, involved six zones in Wuhan. The American scientists were responsible for two of them and for providing devices. The report also recorded contacts with the British and German Environmental Protection Agencies.
11. See the case studies on air pollution in Mun S. Ho and Chris P. Nielsen, eds., *Clearing the Air: The Health and Economic Damages of Air Pollution in China*, vol. 1 (Cambridge, MA: MIT Press, 2007). See also Jun Ma, *The Economics of Air Pollution in China: Achieving Better and Cleaner Growth*, edited by Damien Ma (New York: Columbia University Press, 2017).
12. "Environmental Protection Shifted into Legal Tracks," *WHNJ-EPS* (1988), 97–98.
13. Report on World Environment Day, *WHNJ-EPS* (1988), 98–99.
14. Report on the environmental management of construction projects, *WHNJ* (Wuhan: Wuhan Publishing House, 1988), 98.
15. Report of environmental pollution investigation on four groups of environmental zones in Wuhan, *WHNJ-EPS* (1988), 100.

16 Report on forty township water facilities, *HBNJ-EPS* (1988), 93.
17 Report on the award of national excellent environmental monitoring station to the Wuhan central environmental monitoring station, *WHLJ-EPS* (1988), 102.
18 Reports on Xunshi River and Moshui Lake, *WHLJ-EPS* (1988), 103.
19 Report on Huangjingqiao village's eco-farming, *WHNJ-EPS* (1988), 100–1.
20 Report on Huangxiao River project, Section of Urban-Rural Construction, *WHNJ* (1988), 82.
21 Report on river clearing, Subsection of Dam-Dike Construction, *WHNJ* (1988), 84–85.
22 Annual overview report, *WHNJ-EPS* (Wuhan: Wuhan Publishing House, 1989), 133–34.
23 Noise pollution report on Jianghan Road at Hankou, *WHNJ-EPS* (1989), 135.
24 Report of preliminary effectiveness of air-pollution control, *WHNJ-EPS* (1989), 134.
25 The PRC homepage is at http://www.npc.gov.cn/wxzl/wxzl/2000–12/06/content_4487.htm.
26 Report on the Qiaokou Environmental Protection Court, *WHNJ-EPS* (1989), 134.
27 Report of two pollution cases, *WHNJ-EPS* (1989), 134–35.
28 Reports under the subsections of rural ecology, with the environmental monitoring and research, *WHNJ* (Wuhan: Wuhan Publishing House, 1989), 136–38.
29 Reports for the subsection overview of environmental protection, *WHNJ-EPS* (Wuhan: Wuhan Publishing House, 1990), 106–7.
30 Report of environmental protection education programs, *WHNJ-EPS* (1990), 107.
31 Reports of research institutes, Subsection of Environmental Monitoring and Research, *WHNJ-EPS* (1990), 110.
32 Report on the predicted extinction of the Yangtze dolphin, *WHNJ-EPS* (1990), 109.
33 *Changjiang Daily*, July 9, 2018, http://env.people.com.cn/n1/2018/0709/c1010–30134177.html. The *Daily* claimed that evidence for the extinction of the dolphin was inconclusive.
34 Overview Section, *WHNJ* (Wuhan: Wuhan Publishing House, 1990), 33–34.
35 Report of calling more policy attention to pollution caused by village-owned and town-owned factories, *WHNJ-EPS* (1990), 109.
36 Reports on the improved rural ecology, including the one in the previous note, *WHNJ-EPS* (1990), 108–9.
37 Report on policy experiments with effluent discharge licences, *WHNJ-EPS* (1990), 106–7.
38 Annual climate report, Introduction Section, *WHNJ* (1990), 43–44.
39 Brief reports on local climate changing patterns in 1986–90, Climate Subsection, Overview Section, *WHNJ* (Wuhan: Wuhan Publishing House, 1991), 52–54.
40 Initially published in a bi-monthly journal, *Xiaoshuolin*, in 1991, the novel depicted the daily lives of urban residents who were dealing with the pressures of both tradition and modernization but tenaciously moving on with both sad and joyful moments. This novel was retitled "Hankou Scenery," included in the author's 2014 novel reprint collection.

41 Section of Major Local Events, *WHNJ* (1991), 19–30.
42 Review reports on the previous year's achievements, *WHNJ-EPS* (Wuhan: Wuhan Publishing House, 1991), 119–22.
43 Annual summary report, *WHNJ-EPS* (1991), 119–20.
44 Report on a chemical waste spillover incident at the Wuhan Sulfuric Acid Plant, WHNJ 122.
45 Report on rising cases of dust-related lung disease, *WHNJ-EPS* (1991), 122.
46 Report on the designated funding for pollution control projects, *WHNJ-EPS* (1991), 124.
47 Report on the work of registering effluent discharge accounts, Research and Management Subsection, *WHNJ-EPS* (1991), 122.
48 Update report on the Dong Lake sewage interception project, *WHNJ-EPS* (1991), 125.
49 Report on the establishment of the Tian'ezhou Yangtze Dolphin National Nature Reserve, *WHNJ* (1991), 124.
50 Report on the finless porpoise being officially listed as a Class 1 protected species, *Hubei Daily*, February 10, 2021.

Bibliography

Ash, Robert F., and Richard Louis Edmonds. "China's Land Resources, Environment and Agricultural Production." *China Quarterly* 156 (1998): 836–79.

Avenell, Simon. "From Fearsome Pollution to Fukushima: Environmental Activism and the Nuclear Blind Spot in Contemporary Japan." *Environmental History* 17, 2 (2012): 244–76.

Bao, Maohong. "Environmental History in China." *Environment and History* 10, 4 (2004): 475–99.

–. "The Evolution of Environmental Policy and Its Impact in the People's Republic of China." *Conservation and Society* 4, 1 (2006): 36–54.

Barresi, Paul A. "The Chinese Legal Tradition as a Cultural Constraint on the Westernization of Chinese Environmental Law and Policy: Toward a Chinese Environmental Law and Policy Regime with More Chinese Characteristics." *Pace Environmental Law Review* 30, 3 (2013): 1156–1221.

Bina, Olivia. "Environmental Governance in China: Weakness and Potential from an Environmental Policy Integration Perspective." *China Review* 10, 1 (2010): 207–39.

Brekke, Kjell Arne, Ragnhild Haugli Braaten, and Ole Rogeberg. "Buying the Right to Do Wrong – an Experimental Test of Moral Objections to Trading Emission Permits." *Resource and Energy Economics* 42 (2015): 110–24.

Chen, Yu, Gaihe Yang, Sandra Sweeney, and Yongzhong Feng. "Household Biogas Use in Rural China: A Study of Opportunities and Constraints." *Renewable and Sustainable Energy Reviews* 14, 1 (2010): 545–49.

Cheng, Dongshen. "Huanjing Jingjixue Chutan" [Origins of Environmental Economics]. *Social Sciences in Yunnan* 06 (1982): 38–47.

Cheng, Fugu. "Shengtai Jingjixue Yuanliu" [Origins of Ecological Economics]. *Economic Research Journal* 9 (1983): 43–49.

Cohn, Steven Mark. *Competing Economic Paradigms in China: The Co-evolution of Economic events, Economic Theory and Economics Education, 1976–2016*. Oxfordshire: Routledge, 2017.

Common, Mick. "Economics and the Natural Environment: A Review Article." *Journal of Economic Studies* 25, 1 (1998): 57–73.

Courtney, Chris. "The Dragon King and the 1931 Wuhan Flood: Religious Rumors and Environmental Disasters in Republican China." *Twentieth-Century China* 40, 2 (2015): 83–104.

Cronon, William. *Nature's Metropolis: Chicago and the Great West*. New York: W.W. Norton, 2009.

de la Croix, David, and Axel Gosseries. "The Natalist Bias of Pollution Control." *Journal of Environmental Economics and Management* 63, 2 (2012): 271–87.

Deng, Hongbin, and et al. "Jin 50 Nian Lai Jianghan Hu Qun Shuiyu Yanhua Dingliang Yanjiu" [Quantilative Research on Hanjiang Lake Groups in 50 Years]. *Resources and Environment in the Yangtze Basin* 15, 2 (2006): 244–48.

Desrochers, Pierre. "Did the Invisible Hand Need a Regulatory Glove to Develop a Green Thumb? Some Historical Perspective on Market Incentives, Win-Win Innovations and the Porter Hypothesis." *Environmental and Resource Economics* 41, 4 (2008): 519–39.

Douai, Ali, Andrew Mearman, and Ioana Negru. "Prospects for a Heterodox Economics of the Environment and Sustainability." *Cambridge Journal of Economics* 36, 5 (2012): 1019–32.

Dunstan, Helen. "Official Thinking on Environmental Issues and the State's Environmental Roles in Eighteenth-Century China." In *Sediments of Time: Environment and Society in Chinese History*, ed. Mark Elvin and Curirong Liu, 585–614. Cambridge: Cambridge University Press, 1998.

Economy, Elizabeth. "Environmental Governance: The Emerging Economic Dimension." *Environmental Politics* 15, 2 (2006): 171–89.

—. *The River Runs Black: The Environmental Challenge to China's Future*. Ithaca: Cornell University Press, 2010.

Edmonds, Richard Louis. "The Environment in the People's Republic of China 50 Years On." *China Quarterly* 159 (1999): 640–49.

—. "The Evolution of Environmental Policy in the People's Republic of China." *Journal of Current Chinese Affairs* 40, 3 (2011): 13–35.

—. "Studies on China's Environment." *China Quarterly* 156 (1998): 725–32.

Elvin, Mark. "The Environmental History of China: An Agenda of Ideas." *Asian Studies Review* 14, 2 (1990): 39–53.

Elvin, Mark, and Cuirong Liu, eds. *Sediments of Time: Environment and Society in Chinese History*. Cambridge: Cambridge University Press, 1998.

Faure, Michael G., and Jing Liu. "Compensation for Environmental Damage in China: Theory and Practice." *Pace Environmental Law Review* 31, 1 (2014): 240–321.

Fu, Zhufu. "Youguan Zhongguo Jingjishi De Ruogan Teshu Wenti" [Three Fundamental Questions in Environmental Economics]. *Economic Research Journal* 7 (1978): 49–61.
Gaubatz, Piper. "New Public Space in Urban China." *China Perspectives*, 4 (2008): 73–84.
Geall, Sam. *China and the Environment: The Green Revolution*. London: NBN International, 2013.
Gilley, Bruce. "Legitimacy and Institutional Change: The Case of China," *Comparative Political Studies* 41, 3 (2012): 259–84.
Goldman, Merle, and Andrew Gordon. *Historical Perspectives on Contemporary East Asia*. Cambridge, MA: Harvard University Press, 2000.
Goldstone, Jack A. "The Rise of the West – or Not? A Revision to Socio-Economic History." *Sociological Theory* 18, 2 (2000): 175–94.
Gong, Sheng-sheng. "Historical Variation and Sustainable Utilization of the Jianghan-Dongting Plain's Wetland." *Resources and Environment in the Yangtze Basin* 11, 6 (2002): 569–74.
Grunow, Dieter. "Structures and Logic of EP Implementation and Administration in China." *Journal of Current Chinese Affairs* 40, 3 (2011): 37–75.
Hamrin, Carol Lee, and Timothy Cheek. *China's Establishment Intellectuals*. New York: Routledge, 2023.
Heggelund, Gørild. "Resettlement Programmes and Environmental Capacity in the Three Gorges Dam Project." *Development and Change* 37, 1 (2006): 179–99.
Heilmann, Sebastian. "From Local Experiments to National Policy: The Origins of China's Distinctive Policy Process." *China Journal* 59 (2008): 1–30.
Hesketh, Therese, Xudong Zhou, and Yun Wang. "The End of the One-Child Policy: Lasting Implications for China." *Jama* 314, 24 (2015): 2619–20.
Ho, Mun S., and Chris P. Nielsen, eds. *Clearing the Air: The Health and Economic Damages of Air Pollution in China*. Vol. 1. Cambridge, MA: MIT Press, 2007.
Ho, Peter, ed. *Developmental Dilemmas: Land Reform and Institutional Change in China*. London: Routledge, 2005.
Ho, Peter, and Richard Louis Edmonds. "Perspectives of Time and Change: Rethinking Embedded Environmental Activism in China." *China Information* 21, 2 (2007): 331–44.
Hou, Wenhui. "Reflections on Chinese Traditional Ideas of Nature." *Environmental History* 2, 4 (1997): 482–93.
Howe, Charles W., and Dwight R. Lee. "Priority Pollution Rights: Adapting Pollution Control to a Variable Environment." *Land Economics* 59, 2 (1983): 141–49.
Huang, Jin-liang. "Jin 500 Nian Jianghan Pingyuan Huqu Tudi Kaifa De Lishi Fanshi" [Historical Thought on Land Exploitation in Recent 500 Years in Jianghan Plain]. *Journal of Central China Normal University (Natural Science Edition)* 35, 4 (2001): 485–88.
Huang, Philip, C.C. "Public Sphere/Civil Society in China? The Third Realm between State and Society." *Modern China* 19, 2 (1993): 216–40.

Huang, Xibing, Dingtao Zhao, Colin G. Brown, Yanrui Wu, and Scott A. Waldron. "Environmental Issues and Policy Priorities in China: A Content Analysis of Government Documents." *China: An International Journal* 8, 2 (2010): 220–46.

Hughes, J. Donald. *What Is Environmental History?* Cambridge: Polity, 2006.

—. "The Organization of Environmental Protection in China." *China Quarterly* 156 (1998): 757–87.

Jiang, Yingguang. "Youguan Shengtai Jingji De Jige Lilun Wenti" [Several Theoretical Questions in Ecological Economics]. *Economic Research Journal* 10 (1983): 56–60.

Jin, Junjian. "Jiaqiang Zhongguo Jingjishi Yanjiu Shi Fazhan Jingji Xueke De Yixiang Zhongyao Zhanlüe Renwu" [A Strategic Task to Strengthen the Researches of Chinese Economic History for the Development Economics Studies]. *Economic Research Journal* 10 (1983): 61–68.

Kornai, Janos. "The Soft Budget Constraint." *Kyklos* 39, 1 (1986): 3–30.

Lai, Yu-Bong. "The Optimal Distribution of Pollution Rights in the Presence of Political Distortions." *Environmental and Resource Economics* 36, 3 (2007): 367–88.

Levy, Jason K. "Multiple Criteria Decision Making and Decision Support Systems for Flood Risk Management." *Stochastic Environmental Research and Risk Assessment* 19, 6 (2005): 438–47.

Li, Gang. "Kexuejishu Shi Guanxi Guojia Shengsicunwang De Yige Zhongyao Yinsu—Xuexi Zhongguo Jindai Jingjishi De Yidian Tihui" [Science Is the Important Factor for National Survival – Some Reflections on the Study of Chinese Modern Economic History]. *Economic Research Journal* 5 (1978): 60–64.

Li, Keguo. "Zhongguo De Huanjing Jingji Zhengce" [Chinese Environmental Economy Policy]. *Ecological Economy* 11 (2000): 39–42.

Lin, Justin Yifu, and Guofu Tan. "Policy Burdens, Accountability, and the Soft Budget Constraint." *American Economic Review* 89, 2 (1999): 426–31.

Liu, Guihuan, Jun Wan, Huiyuan Zhang, and Lijie Cai. "Eco-Compensation Policies and Mechanisms in China." *Review of European Community and International Environmental Law* 17, 2 (2008): 234–42.

Liu, Lingxuan, Bing Zhang, and Jun Bi. "Reforming China's Multi-Level Environmental Governance: Lessons from the 11th Five-Year Plan." *Environmental Science and Policy* 21 (2012): 106–11.

Liu, Shihua. "Lun Shengtai Jingji Xuqiu" [Demand of Ecological Economy]. *Economic Research Journal* 4 (1988): 77–79.

Liu, Yun. "Revisiting Hanyeping Company (1889–1908): A Case Study of China's Early Industrialisation and Corporate History." *Business History* 52, 1 (2010): 62–73.

Liu, Zhongyi, and Liu Yaochuang. "Shixian Nongye De Liangge Liangxing Xunhuan" [Materializing Two Benign Cycles of Agriculture]. *Economic Research Journal* 5 (1982): 7–12.

Lu, Jian, and Chris King-Chi Chan. "Collective Identity, Framing and Mobilisation of Environmental Protests in Urban China: A Case Study of Qidong's Protest." *China: An International Journal* 14, 2 (2016): 102–22.

Lu, Yiyi. "Environmental Civil Society and Governance in China." *International Journal of Environmental Studies* 64, 1 (2007): 59–69.

Lyon, Randolph M. "Auctions and Alternative Procedures for Allocating Pollution Rights." *Land Economics* 58, 1 (1982): 16–32.

Ma, Chuandong. "Tigao Nengyuankaifa Liyong De Jingji Shengtaixiaoyi: [Promoting Ecological-Economic Benefit from Upgrading Energy Utilization]. *Economic Research Journal* 2 (1985): 58–62.

Ma, Jun. *The Economics of Air Pollution in China: Achieving Better and Cleaner Growth*, edited by Damien Ma. New York: Columbia University Press, 2017.

Mao, Yushi. "Huanjing Jingjixue Zhong Sange Jiben Wenti" [Three Fundamental Questions in Environmental Economics]. *Economic Research Journal* 7 (1982): 60–63.

Mao, Yushi. "Huanjing Jingjixue Zhong Sange Jiben Wenti" [Three Fundamental Questions in Environmental Economics]. *Economic Research Journal* 7 (1982): 60–63.

–. "Zeyou Fenpei Yuanli Jianjie" [A Short Introduction to the Principles of Optimal Allocation]. *Economic Research Journal* 12 (1980): 65–68.

Marks, Robert. *China: Its Environment and History*. World Social Change. Toronto: Rowman and Littlefield, 2012.

Marks, Robert B. "Why China?" *Environmental History* 10, 1 (2005): 56–58.

Martinez-Alier, Juan. *Ecological Economics: Economics, Environment, and Society*. Oxford: Basil Blackwell, 1987.

–. "The Environment as a Luxury Good or 'Too Poor to Be Green'?" *Ecological Economics* 13, 1 (1995): 1–10.

Melosi, Martin V. *Effluent America: Cities, Industry, Energy, and the Environment* Pittsburgh: University of Pittsburgh Press, 2000.

Meng, Tianguang, Jennifer Pan, and Ping Yang. "Conditional Receptivity to Citizen Participation:Evidence From a Survey Experiment in China." *Comparative Political Studies* 50, 4 (2017): 399–433.

Montgomery, Cynthia A., and Michael E. Porter, eds. *Strategy: Seeking and Securing Competitive Advantage*. Boston: Harvard Business School Press, 1991.

Moser, Adam J., and Tseming Yang. "Environmental Tort Litigation in China." *Environmental Law Reporter* 41, 10 (2011): 10895–901.

Moulik, T.K. "The Biogas Program in India and China." *Ambio* 14, 4–5 (1985): 288–92.

North, Douglass C. *Understanding the Process of Economic Change*. Princeton Economic History of the Western World. Princeton: Princeton University Press, 2005.

Palmer, Michael. "Environmental Regulation in the People's Republic of China: The Face of Domestic Law." *China Quarterly* 156 (1998): 788–808.

Perdue, Peter C. "Zhongguo Huanjingshi Yangjiu Xiangzhuang Ji Qushi" [Trends and status of Chinese environmental history research]." *Jianghan Tribune* 5 (2014): 38–40.

Perry, Elizabeth J. *Anyuan: Mining China's Revolutionary Tradition*. Berkeley: University of California Press, 2012.
Pietz, David Allen. *The Yellow River: The Problem of Water in Modern China*. Cambridge, MA: Harvard University Press, 2015.
Pomeranz, Kenneth. "Chinese Politics and Environmental History." *Environmental History* 12, 2 (2007): 352–54.
–. *The Great Divergence: China, Europe, and the Making of the Modern World Economy*. Princeton, NJ: Princeton University Press, 2009.
–. "The Transformation of China's Environment, 1500–2000." In *The Environment and World History*, ed. Edmund Burke and Kenneth Pomeranz, 118–64. Berkeley: University of California Press, 2009.
Porter, Michael E., and Claas van der Linde. "Toward a New Conception of the Environment-Competitiveness Relationship." *Journal of Economic Perspectives* 9, 4 (1995): 97–118.
Qi, Gao. "Public Interest Litigation in China: Panacea or Placebo for Environmental Protection?" *China: An International Journal* 16, 4 (2018): 47–75.
Qiao, Tongfong. "Jinnianlai Shengtai Jingji Wenti Taolun Zongshu" [an Overview on Ecological – Economic Problems in Recent Years]. *Economic Research Journal* 2 (1982): 77–80.
Rao, Qinzi. "Hubeisheng Hubo Diaocha" [A Survey of Hubei Lakes]. *Chinese Science Bulletin* 10 (1954): 71–83.
Ross, Lester. "China: Environmental Protection, Domestic Policy Trends, Patterns of Participation in Regimes and Compliance with International Norms." *China Quarterly* 156 (1998): 809–35.
Rowe, William T. *Hankow: Commerce and Society in a Chinese City, 1796–1889*. Vol. 1. Stanford: Stanford University Press, 1984.
–. *Hankow: Conflict and Community in a Chinese City, 1796–1895*. Stanford: Stanford University Press, 1989.
Ryan, Donald R. "Transferable Discharge Permits and the Control of Stationary Source Air Pollution: Reply." *Land Economics* 59, 1 (1983): 126–27.
Schmalzer, Sigrid. *Red Revolution, Green Revolution: Scientific Farming in Socialist China*. Chicago: University of Chicago Press, 2016.
Shapiro, Judith. *China's Environmental Challenges*. Malden, MA: Polity Press, 2012.
–. *Mao's War against Nature: Politics and the Environment in Revolutionary China*. Cambridge, MA: Cambridge University Press, 2001.
Shen, Manhong. "Lun Huanjing Jingji Shouduan" [Exploring the Market – Based Instruments for an Economic – Environmental Policy Framework]. *Economic Research Journal* 10 (1997): 54–61.
Shi, Tian. "Ecological Economics in China: Origins, Dilemmas and Prospects." *Ecological Economics* 41, 1 (2002): 5–20.
Sitaraman, Srini. "Regulating the Belching Dragon: Rule of Law, Politics of Enforcement, and Pollution Prevention in Post-Mao Industrial China." *Colorado Journal of International Environmental Law and Policy* 18, 2 (2007): 267–336.

—. "Regulating the Environment: Assessing China's Domestic Environmental Law and Participation in International Treaties." *China Review* 6, 1 (2006): 183–96.

Skinner, Mark W., Alun E. Joseph, and Richard G. Kuhn. "Social and Environmental Regulation in Rural China: Bringing the Changing Role of Local Government into Focus." *Geoforum* 34, 2 (2003): 267–81.

Smil, Vaclav. *The Bad Earth: Environmental Degradation in China*. Armonk, NY/London: M.E. Sharpe/Zed Press, 1984.

—. "China's Energy and Resource Uses: Continuity and Change." *China Quarterly* 156 (1998): 935–51.

Songster, E. Elena. *Panda Nation: The Construction and Conservation of China's Modern Icon*. Oxford: Oxford University Press, 2018.

Spash, Clive L., and Anthony Ryan. "Economic Schools of Thought on the Environment: Investigating Unity and Division." *Cambridge Journal of Economics* 36, 5 (2012): 1091–1121.

Spence, Jonathan D. *The Search for Modern China*. 1st edition. New York: Norton, 1990.

Stern, Rachel E. *Environmental Litigation in China: A Study in Political Ambivalence*. Cambridge: Cambridge University Press, 2013.

—. "From Dispute to Decision: Suing Polluters in China." *China Quarterly* 206 (2011): 294–312.

—. "On the Frontlines: Making Decisions in Chinese Civil Environmental Lawsuits." *Law and Policy* 32, 1 (2010): 79–103.

Sun, Zhonglian. "Kuangchanziyuan De Youchang Shiyong" [Compensation Costs in Using Mineral Resources]. *Economic Research Journal* 2 (1986): 70–75.

Swanson, Kate E., Richard G. Kuhn, and Wei Xu. "Environmental Policy Implementation in Rural China: A Case Study of Yuhang, Zhejiang." *Environmental Management* 27, 4 (2001): 481–91.

Tang, Qixiang. "Yumeng Yu Yumeng Ze" [Yunmengg and Yumeng Mash]. *Fudan Journal (Social Sciences Edition)* 1 (1980): 1–11.

Thompson, E.P. "The Moral Economy of the English Crowd in the Eighteenth Century." *Past and Present* 50 (1971): 76–136.

Tilt, Bryan. "Industrial Pollution and Environmental Health in Rural China: Risk, Uncertainty and Individualization." *China Quarterly* 214 (2013): 283–301.

—. *The Struggle for Sustainability in Rural China: Environmental Values and Civil Society*. New York: Columbia University Press, 2010.

Van Buren, E. Ariane. "Biogas beyond China: First International Training Program for Developing Countries." *Ambio* 9, 1 (1980): 10–15.

Vermeer, Eduard B. "Industrial Pollution in China and Remedial Policies." *China Quarterly* 156 (1998): 952–85.

Wang, Shaoguang. *Failure of Charisma: The Cultural Revolution in Wuhan*. Oxford: Oxford University Press, 1995.

Whyte, Martin King, Wang Feng, and Yong Cai. "Challenging Myths about China's One-Child Policy." *China Journal* 74 (2015): 144–59.

Worster, Donald. *Wealth of Nature: Environmental History and the Ecological Imagination.* New York, Oxford: Oxford University Press, 1994.

Xia, Zuzhang. *Domestic Biogas in a Changing China: Can Biogas Still Meet the Energy Needs of China's Rural Households?* London: International Institute for Environmental and Development, 2013.

Xian, Ming, and Lu Bing. "Maoyushi De Jingjixue Rensheng" [Mao Yushi'S Life in Economics]. *China Reform* 11 (2013): 37–42.

Xin, Gu. "Plural Institutionalism and the Emergence of Intellectual Public Spaces in Contemporary China: Four Relational Patterns and Four Organizational Forms." *Journal of Contemporary China* 7, 18 (1998): 271–301.

Xu, Dixin. "Shixian Sihua Yu Shengtai Jingjixue" [Four Modernizations and Ecological Economics]. *Economic Research Journal* 11 (1980): 14–18.

—. "Youguan Nongye Jingji De Jige Wenti" [Several Problems in Agricultural Economics]. *Economic Research Journal* 12 (1978): 8–16.

Xu, Xinchuang, Quansheng Ge, Jingyun Zheng, and Chenwu Liu. "Hubei sheng jin 500 nian quyu ganshi xulie chongjian jiqi bijiao fenxi" [Hube's 500 years of Dry versus Wet Weather Re-modeling and Comparative Analysis]. *Geography Research* 29, 6 (2010): 1045–55.

Xue, Qinghao. "Zhongguokexueyuan Zucheng Hubo Diaochadui Qu Hubeisheng Yanjiu Yulei Yangzhi Wenti" [Fishery Questions of Hubei Investigated by the Lake Team Sent by the China Institute of Science]." *Chinese Science Bulletin* 7 (1953): 97–98.

Yang, Tseming. "Mysteries, Myths, and Misunderstandings." *Environmental Forum* 33, 2 (2016): 36–42.

Yang, Zhongping. "Yao Jiaqiang Dui Jingjishi De Yanjiu" [Strengthening the Researches on Economic History]. *Economic Research Journal* 10 (1981): 19–20.

Yeh, Emily T. "The Politics of Conservation in Contemporary Rural China." *Journal of Peasant Studies* 40, 6 (2013): 1165–88.

Yeh, Emily T., Kevin J. O'Brien, and Jingzhong Ye. "Rural Politics in Contemporary China." *Journal of Peasant Studies* 40, 6 (2013): 915–28.

Yin, Hongfu, Guangrun Liu, Jiangao Pi, Guojin Chen, and Changan Li. "On the River–Lake Relationship of the Middle Yangtze Reaches." *Geomorphology* 85, 3 (2007): 197–207.

Zeng, Yi, and Therese Hesketh. "The Effects of China's Universal Two-Child Policy." *The Lancet* 388, 10054 (2016): 1930–38.

Zhang, Bo, and Cong Cao. "Four Gaps in China's New Environmental Law: Implementation and Accountability Will Remain Challenging, Especially at the Local Level, Warn Bo Zhang and Cong Cao." *Nature* 517, 7535 (2015): 433–35.

Zhang, Chaozhun. "Guanyu Zhongguoshi Xiandaihua Daolu De Tantao" [Exploring the Chinese Modernization Path]. *Economic Research Journal* 9 (1979): 3–9.

Zhang, Jianming. "Qingdai Jianghan — Dongtinghuqu Diyuan Nongtian De Fazhan Jiqi Zonghe Kaocha" [a Comprehensive Analysis on the Dyke-Farm in the Han

River-Dongting Lake Area in Qing-China]. *Chinese Aggriculture History* 2 (1987): 72–88.
Zhang, Jiayan. *Coping with Calamity: Environmental Change and Peasant Response in Central China, 1736–1949*. Vancouver: UBC Press, 2014.
Zhang, Lei, Arthur P.J. Mol, and David A. Sonnenfeld. "The Interpretation of Ecological Modernisation in China." *Environmental Politics* 16, 4 (2007): 659–68.
Zhang, Xiao. "Zhongguo Huanjing Zhengce De Zongti Pingjia" [An Overall Assessment on China'S Environmental Policy]. *Social Sciences in China* 3 (1999): 88–99.
Zhang, Xuehua. "China's Environmental Administrative Enforcement System." *Environmental Law Reporter: News and Analysis* 41, 10 (2011): 10890–94.
–. "Implementation of Pollution Control Targets in China: Has a Centralized Enforcement Approach Worked?" *China Quarterly* 231 (2017): 749–74.
Zhang, Xuehua, Leonard Ortolano, and Zhongmei Lu. "Agency Empowerment through the Administrative Litigation Law: Court Enforcement of Pollution Levies in Hubei Province." *China Quarterly* 202 (2010): 307–26.
Zhao, Yuhong. "Environmental Dispute Resolution in China." *Journal of Environmental Law* 16, 2 (2004): 157–92.
Zhuang, Renan, and Won W. Koo. "The Role of the Agricultural and Industrial Sectors in China's Economic Growth: Are They Twin Brothers?" *China: An International Journal* 6, 2 (2008): 299–315.

Index

Notes: "(f)" after a page number indicates a figure; "(t)" after a page number indicates a table. Naming discrepancies exist among some local government titles, which had different titles at various stages and sometimes simultaneously. In addition, the translation of titles resulted in some nuanced differences in their informal usage in various local contexts.

agriculture: biodiversity, 150–51; chemical fertilizers and pesticides, 77; climate reports, 156, 191*n*39; cropland contamination, 75–77; DDT ban, 77; eco-farming, 150; vs industrial interests, 52–53, 54, 55, 58, 129, 176*n*53; inter-agency collaboration, 66, 98–99; rice paddies contamination, 72–74, 180*n*39. *See also* soil pollution

air pollution: acid rain, 156; administrative measures, 63–64; chemical dust effects, 66, 179*n*25; civilian petitions, 63–65, 66, 179*n*24; containment devices, 65–66; dust-control and monitoring, 69–70, 143, 149, 152, 153, 190*n*8; dust-related lung disease (workplace safety), 65, 157, 192*n*45; environmental history, 60; factory shutdown orders, 64, 68; factory workers complaints, 68–70; field surveys, 63; fluoride dust, 67; health risks, 65, 157, 192*n*45; legislation enforcement, 153; pollution-control devices, 63; pollution-control investment projects, 65–66, 179*nn*21–22; public letters of appeal, 66–68; records, 10; sulphuric acid factory, 69; university faculty complaints, 66–68, 179*n*29. *See also* industrial waste emissions

Anti-rightest Campaign, 95

Atmospheric Pollution Prevention and Control Law, 152

Baiji Dolphin House, 159(f)

Bao, Maohong, 14

Barresi, Paul, 16

Beijing poverty, 137

Bio-energy Utilization Research Team, 95
biogas economy: administrative directives, 96–97; conferences, 93, 96, 97; digester device research, 95; digester restoration, 96; digester units statistics, 93–94, 183*n*45; industrial microwave, 96; innovation and technology, 93; labour and materials report, 95–96, 184*n*52; local vs state policy, 94, 96–97, 103; mass movement, 86, 94, 97, 103, 165–66; offices, 86–87, 92–97, 98; promotion, 98; reports, 95–96; research institutes, 95; research proposals (People's Letter), 94–95; service station network, 97; state endorsement, 96; training programs, 96–97
Biogas Promotion Movement, 86, 103
birds (storks), 150
Biwu Dong (1886–1975), 95–96, 97, 98
Boxer Rebellion (circa 1900–01), 31

CASS. *See* Chinese Academy of Social Sciences (CASS)
CEES. *See* Chinese Ecological Economics Society (CEES)
CCP. *See* Chinese Communist Party (CCP)
Changjiang (Yangtze/Yangzi) River: five-tranche categorization system, 152; maps, 41(f), 47(f); porpoises, 158, 192*n*50; railway bridge, 20; reservoir, 46; river dolphin decline, 154–55; river silting, 22–23; wastewater monitoring, 149
Changjiang (Yangtze) Water Resources Commission, 34
Cheek, Timothy, 164
chemical industry: air pollution, 66, 68–70; ammonia leaks, 128–29; chemical dust effects, 66–67, 179*n*25; effluent pollution reports, 50, 52–53, 90, 128–29, 153, 176*nn*50–51, 197*n*46; factory workers protests, 69; loss-compensation negotiations, 54; pollution-detection devices, 53–54; waste-discharge penalties, 52–54, 153. *See also* industrial waste emissions
Chen, Mingzhi, 111
Chen, Muhua, 53, 176*n*55
Chen, Pixian, 63, 178*n*10
Cheng, Fugu, 109
China. *See* People's Republic of China (PRC)
China Association of Environmental Protection Industry, 140–41
China Ministry of Water Resources, 34
China Society for Environmental Science (CSES), 146
China Society of Territorial Economists (CSTE), 109
China University of Geosciences (Wuhan), 79
Chinese Academy of Science, 79, 95, 97, 155
Chinese Academy of Social Sciences (CASS), 109, 111, 164
Chinese Communist Party (CCP): central committee, 81; chairman, 19–20, 41, 79–80, 81, 96, 97, 116, 129; college-entry examinations, 79–80, 180*n*51; education reform, 79–81, 180*n*51, 180*n*57; model experiments, 85, 93, 183*n*41; municipal committee, 69; provincial committee, 63, 178*n*10, 180*n*39, 187*n*23; provincial revolutionary committee, 56–57, 89, 95, 96–97, 123; vice-chairman, 74, 95–96
Chinese Ecological Economics Society (CEES), 108–9
Chinese People's Political Consultative Conference, 146, 180*n*44
chronicles (*nianjian*) and yearbooks, 6
civil (public) service, 48, 78, 84, 99, 180*n*51
CNN report, 128–29

Index

Confucian exam (*keju*), 78, 81
construction projects, 94, 136, 146–47, 151
County Ammonia Plant, 128
Cultural Revolution, 34, 79, 88, 105, 163

Dawu county: about, 61, 74–75; environmental protection office, 75; map, 76(f); revolutionary legacy, 74–75
Daye Company/Group (Daye Iron and Steel Company/Daye Special Steel Company), 50, 71–72, 74, 100, 101, 124, 185*n*72
Daye county: about, 61; administrative status, 72; emissions violations letter, 124, 189*n*49; industrial soil pollution, 71–74, 180*n*37; map, 73(f)
Daye Lake, 48
Daye Ore Mining Company, 28, 30, 39
de la Croix, David, 114
Deng Xiaoping, 79–80, 81, 116
Desrochers, Pierre, 122
Development Economics, 105
dolphin (*baiji*), 154–55, 158–59(f), 191*n*33
Dong (Donghu/East) Lake: about, 40–41; effluent discharge evidence, 44–45, 89, 91, 175*n*24; effluent discharge licences, 156; effluent discharge sources, 43–44, 175*n*15; factories on, 43–44; five-tranche categorization system, 152; invasive plants (water hyacinth), 44; local vs state jurisdiction over, 44; map, 41(f); military map facility on, 43; pollution research project, 142–43; sewage interception project, 136, 150, 157–58; sunrise and sailing on, 42(f); swimming, 44; universities on, 43, 45–46, 174*n*14; zone restrictions, 43
Donghu (East Lake) Sewage Interception and Treatment Project, 136
Donghu Water Facility, 143–44
Donghu Water Source Protection Initiative, 136, 137

Dongting Lake, 19, 22, 23
Douai, Ali, 118
DPMS. *See* Wuhan Municipal Disease Prevention and Monitoring Station (DPMS)
Dynasties of China, 21(t), 23, 58, 78

EARs. *See* environmental assessment reports (EARs)
ecological economics, 108–10, 185*n*2
Economic Research Journal ([ERJ] Jingji Yanjiu), 110–11, 119
Economy, Elizabeth, 17, 85
Edmonds, Richard, 18
education: college-entry examinations, 79–80, 180*n*51; imperial officer examinations, 78, 81; industrialization, 80–81, 181*n*57; internalization and modernization, 80; *keju* exam system, 78, 81, 180*n*51; reform, 79–81, 180*n*51, 180*n*57. *See also* universities
effluent pollution: ammonia leaks, 128–29; anti-pollution investments, 147; case studies, 7, 36, 39, 40–55; chemical industry, 50, 52–53, 90, 128–29, 153, 176*nn*50–51, 197*n*46; containment directives, 40, 66, 106, 174*n*10; covert waste-dumping, 48; dike containment, 49; discharge licences, 156; emergency notices, 40, 56, 89–90, 128, 189*n*43; emissions statistics, 142; *ex-ante* vs *ex-post* response, 40; fisheries contamination, 4, 40, 49, 90, 146, 174*n*10; government memoranda on, 40, 44, 175*n*19, 175*n*22; government suppression, 128–29; media response, 44, 128–29; military arsenal effluent, 54, 177*n*58; policy priorities, 36–37, 174*n*4; records on, 5, 10; reports, 40, 48, 89–90, 157; residents reaction to, 44–45; saline discharge sources, 50, 52–53; sewage interception and treatment, 136–37, 149–50,

157–58, 189n46; under-regulation, 40; victim compensation, 146; waste-discharge fees and refund policy, 145, 152, 153, 157; water-supply contamination, 44–45, 52, 74, 128–29; wastewater monitoring, 149–50. *See also* industrial waste emissions; lakes and rivers

Elvin, Mark, 14

environmental assessment reports (EARs), 146–47

environmental awareness: biodiversity, 150–51; biogas movement, 86, 94, 97, 103, 165–66; campaigns, 120–21; civilian petitions, 63, 64–65, 66, 179n24; conferences, 120; eco-farming, 150; ecological diversity, 150; environmental history, 33–35; government memoranda, 40; ignorance towards, 7; journals, 34–35; letters of appeal, 66, 67–71, 77, 146, 180n50; media response, 48, 175n33; protests, 63, 64–70, 74, 77–78, 146; public pressure and political will, 161; rural vs urban, 70–71, 82; scholarship, 34–35; strikes, 69; unrest, 4; water-supply safety, 44–45, 175n24; whistle-blowers, 48; workplace safety, 40

environmental governance and protection policy: about, viii–x; administrative directives, 71, 80, 89, 112, 124–25, 189n51; administrative vs judicial processes, 15–16; administrative vs jurisdictional boundaries, 22–23; administrative reforms, 18; authority limitations, 57–59; beginnings, 163, 165; conferences, 63, 72, 74, 91, 96, 97, 120, 153, 183n29; cultural factors, 8–9; decentralization, 16, 165; developmental dilemma, 13; enforcement by litigation, 59, 177n68; events timeline, 131–32(t); game-theory analysis, 115; getting-the-prices-right principle, 13; global regulation, 104; historical analysis, 3–5, 7; implementation reports, 155–56; intra-sectoral conflict, 104–5, 112, 123, 185n1, 186n12; legislation, 15–17, 88, 92, 107, 153–54, 157, 182n17; local vs state model experiments, 85, 93, 183n41; local vs state priorities, ix–x, 8–11, 37, 44, 82, 85–87, 162, 166–68, 182n6, 182n10; methodologies and theories, 13, 83, 85, 115, 162; moral economy, 114, 164; neutral interpretations, 4–5, 102, 103, 162–63; non-linear evolutions, 36, 174n2; performance evaluation review, 153–54; permission to pollute, 55, 114, 124; planning committees, 87, 100, 112; plural institutionalism, 13; policy reports, 71, 97, 157; political impacts, 17, 163–64; political-economic vs eco-environmental, 55–58, 71, 77; pollution rights (right to pollute), 4, 7, 8, 114–15, 120, 126–28, 130, 163, 166, 167; private vs public space, 13; proceeding from point to surface (*youdian daomian*), 85; proposals, 39, 94, 96, 140, 143; regulatory failures, 7–8, 9, 10, 57–59, 83, 100–103, 113, 120, 122, 163–67, 188n42; research institutes, 33–34, 92, 95, 147, 154; shift, 13, 16, 18, 55, 94, 121, 147, 167–68; symbiotic relations, x, 83; symposium, 53, 176n55; ten regulations, 113; themes, 4, 13, 14, 87; three industrial waste governance movement, 86, 88–89; two events/same time fallacy (*post hoc ergo propter hoc*), 162; victim compensation, 55–56, 177n62; waste-discharge fees and refund policy, 139, 144–45, 146, 152, 153, 155, 157; without fury/bitterness and bias/partiality (*sine ira et studio*) approach, 83. *See also* environmental-economic policy environmental history: biogas movement, 86; ecological and environmental economics, 108–11, 185n2; environmental awareness, 33–35; flood diversion projects, 23–25(f), 26, 34, 172n46; flood incidents, 20–21(t), 22; governance beginnings, 163, 165; vs industrial heritage, 27–33; mass movements (*qunzhong yundong*), 85–86, 88–89, 103, 182n5; New West approach, 15; non-governmental organizations (NGOs), 164–65; overview, 161; revisionist approach, 38, 174n5; river silting, 22–23; scholarship, 13–19; wetlands, 22

environmental protection: agricultural vs industrial interests, 52–53, 54, 55, 58, 129, 176n53; construction projects, 94, 136, 146–47, 151; degradation, 87; disease prevention and monitoring, 90, 91; education centre, 154; legislation, 15–17, 88, 92, 107, 153–54, 157, 182n17; newspaper reports, 135; noise control devices, 140–41(f); performance scores, 154; pollution control and detection devices, 48–49, 53–55, 139–41(f), 142, 176n55; pollution reports, 50, 52–54, 85–86, 90, 128–29, 153, 176nn50–51, 197n46; porpoises, 158–59(f), 192n50; research exchange collaboration, 144, 190n10; river dolphin (*baiji*), 154–55, 158–59(f), 191n33; rural vs urban, 55–56, 70–71, 82; state receptivity, 81, 181n60; ten regulations, 113; training programs, 96–97, 121, 154; water-quality monitoring, 48, 106, 140, 142–44, 149–50, 152, 191n17; Western technology, 188n38

environmental protection agencies: administrative notices, 121–22; agricultural agencies, 98; anti-pollution campaigns, 112, 115, 146; beginnings, 84–87, 165, 166; bureau (*ju*) vs department or ministry (*ting*), 99, 184n62; bureau directors conference, 153; bureaucratic weaknesses, 9, 100–3, 107, 113, 122, 127, 163–67, 188n42; definition, 86; emergency directives, 40, 56, 89–90; emissions control reimbursement, 64, 101, 185n72; emissions surveys, 142; environmental assessment reports (EARs), 146–47; feedback, 49; funding, 9, 69, 87, 103, 121, 123, 166, 167; historical analysis, 8; inter-agency relations, 92, 98–99; interference, 167; legal action, 122, 146; local vs national, 100–3; monitoring stations, 106, 123–24, 149, 191n17; negligence, 9, 102; pollution abatement measures, 85–86, 87, 101, 106, 122, 125; public opinion, 167; reforms and restructuring, 98–100; regulation measures, 112–13, 121–22; remediation activities, 165; structure, 99–110, 184n64; training workshops, 121; warning notices, 112, 124–25; as witnesses and policy-solution providers, 161–62

Environmental Protection Court, 153

environmental protection monitoring stations (EPMS), 88, 90–92, 98, 123–24

environmental-economic policy: academic communities, 108–9; academic vs political debate, 119–20; appropriation doctrine, 126; cost-benefit analysis, 10, 57, 70, 95, 105, 106, 131; vs ecological economics, 108–11, 185n2; ecological modernization, 129; economic growth measures, 81–83, 105–6, 112–15, 117, 145; economic rationality, 108, 113; economists role, 107–15; emission-trading schemes, 114; emissions-permit program, 127; environmental justice, 118; evolutions, 37–38; *ex-ante* vs *ex-post* response, 40; gadfly writers, 164; institutional economics,

38, 125, 127; moral economy, 114; political influence, 127–28; political-economic vs eco-environmental, 55–58, 71, 77; pollution abatement investment funding, 100, 113, 125, 144, 189n52; pollution coefficients model, 116–17; private property rights approach, 122; quantitative approach, 115–17, 120, 153, 154; right to pollute (pollution rights), 4, 7, 8, 114–15, 120, 126–28, 130, 163, 166, 167; scholarship, 107–15, 166, 185n2; state vs local priorities, 122–23, 127–30; sustainability, 118; ten regulations, 113; territorial economics, 109, 111; theoretical models, 119–20, 127; tradable emissions permits, 128; trade-offs, 81–83, 105–6, 117, 145; Western scholarship, 117–18, 119, 126. *See also* environmental governance and protection policy

EPMS. *See* environmental protection monitoring stations (EPMS)

Erchen county, 90

ERJ. *See Economic Research Journal ([ERJ] Jingji Yanjiu)*

Eurocentrism, 38, 174n5

Ezhou, 27, 46–47(f), 48, 49, 50

Ezhou Municipal Environmental Protection Bureau, 48

factories: dust-control zones, 152; effluent discharge, 43–44, 71–77, 175n15, 180n37; effluent discharge licences, 156; emissions statistics, 142; energy consumption, 142; noise pollution, 145; occupational hazards, 90; pollution abatement investment funding, 70, 125, 189n52; pollution-free status, 152, 157; warnings, 100, 112, 124–25; waste-discharge fees, 139, 144–45, 146, 152, 155, 157; workplace safety, 157, 192n45. *See also* air pollution; effluent pollution; industrial waste emissions

farming. *See* agriculture

Faure, Michael, 17

fisheries: ammonia contamination, 128–29; chemical dust effects, 66, 179n25; compensation, 49, 72; effluent contamination, 4, 40, 49, 90, 146, 174n10; food poisoning, 49, 128. *See also* lakes and rivers

flood diversion projects, 23–25(f), 26, 34, 172n46

floods, 20–21(t), 22, 25(f)

Four Modernizations, 80, 111, 124, 187n29

Fu (Fuhe/Yun) River: about, 51–52; ammonia leaks, 128–29; five-tranche categorization system, 152; industrial effluent reports, 5, 52–55, 140, 176nn50–51; maps, 41(f), 51(f); military plants, 50, 54; pollution-detection devices, 54; provincial jurisdiction, 52; saline discharge sources, 50, 52–53

gazetteers (*fangzhi*), 5–6
Goldman, Merle, 164
Gongying Zheng, 30, 31
Gordon, Andrew, 164
Gosseries, Axel, 114
Great Leap Forward Movement, 95, 163
Grunow, Dieter, 100
Guangming Daily newspaper, 180n56

Hamrin, Carol Lee, 164
Han Ningfu, 72, 180n39
Han River, 22, 62, 149, 152, 171n41
Hankou district: about, 27, 37, 61, 62, 64; aerial view, 133(f); air pollution, 64–66, 148, 178n16; commercial legacy, 62, 136, 178n5; mega-speakers noise levels, 152–53, 191n23; place-name change (Xiakou), 62; river re-modification project, 136–37, 138(f), 139, 190nn2–3; water-supply safety, 148

Hanyang Arsenal, 28

Hanyang industrial centre, 27, 37, 61–62, 133(f), 134, 140
Hanyang Iron Plant, 27, 28–31, 39
Hanyeping Company: assets, 100; beginnings, 27–28; effluent discharge, 50, 71–72; embezzlement (secret reports), 30–31; foreign investment, 31; management stages, 28; mergers, 39; mismanagement, 30–31, 33, 173n70; soil pollution, 71–72
Heilmann, Sebastian, 85
HEP. *See* Hubei Provincial Bureau of Environmental Protection (HEP)
Ho, Peter, 13
household waste, 40, 43, 104, 129, 136, 150
Howe, Charles, 126
HPA. *See* Hubei Provincial Archives (HPA) Huai River, 17
Huan River, 128
Huang, Philip, 13
Huangjingqiao village, 150, 191n19
Huangpi county, 136, 156
Huangshi city, 46, 50, 72, 73(f), 92
Huangxiao River Environmental Remodification Project, 136–37, 138(f), 139, 150–51, 190nn2–3
Huashi No. 1 Middle High School, 67
Huazhong Agricultural Research Institute, 95
Huazhong Normal University, 67, 79
Huazhong University of Agriculture, 66, 79
Huazhong University of Science, 79, 105
Hubei Academy of Agricultural Science, 66
Hubei Committee of Atmospheric Research, 121
Hubei Construction Committee, 89, 91, 99, 123
Hubei Environmental Protection Committee, 121
Hubei Institute of Hydrogeology and Engineering Geology, 147

Hubei Meteorological Service, 147
Hubei province: about, 4, 12–13, 19; administrative boundaries, 22–23; biogas economy research, 94–95; climate, 19, 20; environmental history, 23, 33–35, 85–87; environmental protection (intra-sectoral conflict), 104–5, 112, 123, 185n1, 186n12; flood diversion project, 23–25(f), 26, 172n46; flood incidents, 20–21(t), 22; forest district, 9; geography, 20; grain production, 26; hydro-electric dam projects, 20, 34; lake place names (*kou*), 23; lakes statistics, 26, 172n52, 172n54; maps, 41(f), 73(f), 76(f); name meaning, 19; as nine-headed bird, 135; place-name changes, 61–62; poem about, 20; population, 26; river-lake systems, 22–23, 171n41; soil pollution, 70–77; turquoise mining, 58; universities, 43, 45–46, 174n14, 175n28; water supply, 45–46, 143–44, 175n28; wetlands, 3, 22–23, 26, 136; wildlife field studies, 9
Hubei Provincial Archives (HPA), 5–6, 39, 56, 174n7, 174n10
Hubei Provincial Bureau of Environmental Protection (HEP), about, 39, 86; administrative notices, 121–22; agricultural vs industrial interests, 52–53, 54, 55, 58, 176n53; air-pollution complainants inquiries, 68–69, 179n32; air-pollution reports, 62–63, 178n8, 178n17; bureau (*ju*) vs department/ministry (*ting*), 99; economic vs political interests, 115; enforcement, 54–55, 69, 122, 153; failures and shortcomings, 7, 52, 57–59, 82–83, 106; funding status, 121; legal action, 101, 122, 153; liaison role, 166; marginalization, 106; monitoring stations, 73, 106, 123–24, 166; negotiated symbiosis, 83; origin story, 88–92, 123,

Index

189n45; political-economic vs eco-environmental goals, 55–58; pollution reports, 54–56; rice paddies contamination reports, 72–74; sanctions and shutdowns, 69; victim compensation, 55–56, 146, 177n62; water-quality monitoring, 48; as witnesses and policy-solution providers, 161–62. *See also* Hubei Provincial Leadership Office of Environmental Protection
Hubei Provincial Bureau of Industries, 95, 99, 184n48
Hubei Provincial Bureau of Metallurgy, 101, 121
Hubei Provincial Bureau of Petroleum and Chemical Products, 69, 70
Hubei Provincial Bureau of Water Utility and Irrigation, 40, 89–90
Hubei Provincial Bureau/Department of Health, 40, 87, 90, 123
Hubei Provincial Committee of the CCP, 18, 63, 178n10, 180n39, 187n23
Hubei Provincial Committee of Science/Provincial Committee of Science (now Hubei Provincial Department of Science and Technology), 34, 97, 142–43
Hubei Provincial Department of Agriculture, 98–99
Hubei Provincial Department of Forestry, 99
Hubei Provincial Department of Science and Technology (formerly Hubei Provincial Committee of Science), 44–45, 48
Hubei Provincial Institute of Environmental Monitoring, 98
Hubei Provincial Leadership Office of Biogas Construction (Hubei Biogas Office/Hubei Provincial Rural Energy Office/Provincial Office of Rural Energy), 86–87, 92–97, 98, 99, 183n45, 184nn48–49

Hubei Provincial Leadership Office of Environmental Protection, 88, 91, 92, 123. *See also* Hubei Provincial Bureau of Environmental Protection (HEP)
Hubei Provincial Leadership Office of Three Industrial Waste Control (Hubei provincial Sanfei Office/Hubei Sanfei Office/Sanfei Office), 89, 90–91. *See also* Hubei Provincial Bureau of Environmental Protection (HEP)
Hubei Provincial Planning Committee, 87, 100
Hubei Provincial Revolutionary Committee, 89, 95, 96
Hubei Provincial Supreme Court, 153
Hubei Research Institute of Environmental Protection, 92
Hubei Revolutionary Committee, 56, 123
Hubei Shuanghuan Science and Technology Corporation, 129
Hubei University, 179n27. *See also* Zhongnan University of Law and Economics (Hubei University)
Hughes, J. Donald, 14
Hunan province, 19, 22, 23–24, 66
hydro-electric dam projects, 20, 34

industrial waste emissions: by-products recycling, 122; emergency notices, 56, 89, 90; emission codes, 106, 112, 187n16; emission-trading schemes (morality experiment), 114; emissions "accidents," 89–90, 128–29, 153, 192n44; fees and fines, 49, 52, 64; global regulation, 104; government memoranda on, 40, 44, 175n19, 175n22; government suppression, 128–29; local reports about, 76–77; monitoring, 90–92; pollutant reduction equipment, 141–42; pollution rights, 114–15, 120, 124, 130; pollution-control reimbursement, 101, 185*n*72; regulatory incentives, 120; statistics, 142; ten regulations, 113; three waste governance movement, 86, 88–89; tradable emission permits, 128; treatment plants, 136–37; types, 88; underregulation, 40; warnings, 100, 112, 124–25. *See also* air pollution; effluent pollution; factories

institutional theory, 37–38

iron and steel industry: beginnings, 27–33; call for investment (*zhaoshang batiao*), 30; construction costs, 29, 173*n*65; embezzlement (secret reports), 30–31; factories, 32(f); factory site selection, 28–29; financial reporting system, 31; foreign investment, 31, 33; investment in, 29, 31, 33, 173*n*65; management stages, 28; merchant management (*guandu shangban*), 28, 29–30; mergers, 28, 39, 172*n*56, 175*n*17; mismanagement, 28–30; political impacts, 31, 33; pollution-control reimbursement, 101, 185*n*72; soil pollution, 71–77, 180*n*37; waste emissions, 40, 46, 48–50, 71–77

Jahiel, Abigail, 17–18, 99–110, 184*n*64
Japan: biogas research, 95, 184*n*49; foreign investment, 31
Jian Han Chemical Factory, 50
Jianghan Plain, 3, 15, 22, 23, 51
Jiangxia borough, 61
Jiangxia Gedian Chemical Factory, 147
Jingjiang Flood Diversion Project (*Jingjiang fenhong gongcheng*), 23–25(f), 26, 34, 172*n*46
Jingjiang (Jing) River, 22, 23–25(f), 26, 29, 172*n*46
Jingxin Academy, 78

Lai, Yu-Bong, 127–28

lakes and rivers: administration and jurisdictional boundaries, 22–23; ammonia leaks into, 128–29; dike-farm, 171*n*41; five-tranche categorization system, 152; flood incidents, 20–21(t), 22, 25(f), 137; imaginary world (*jianghu*), 84; place names (*kou*), 23; porpoises, 158–59(f), 192*n*50; river dolphin (*baiji*), 154–55, 158–59(f), 191*n*33; river silting, 22–23, 46, 136–37, 151; saline discharge sources, 50, 52–53; statistics, 26, 172*n*52, 172*n*54; urban, 46, 140, 145; water-quality monitoring, 48, 106, 140, 142–43, 149–50, 152, 191*n*17; watercourse modification project, 136–37, 138(f), 139, 151, 190*nn*2–3. *See also* effluent pollution; fisheries; *specific names of lakes and rivers*

Lao She (Colin C. Shu or Lau Shaw, Qingchun Shu), 137
Leadership Office of Environmental Protection. *See* Hubei Provincial Leadership Office of Environmental Protection; National Leadership Office of Environmental Protection
Learn from Dazhai in Agriculture Movement (Nongye xue dazai), 97
Lee, Dwight, 126
legends and folklore, 9, 22
Leontief input-output matrix model, 116
Li Ci, 156, 191*n*40
Li Renzi, 69
Li Xiannian, 23, 74, 180*n*44
Lianghu Academy, 78
Liangzi Lake, 46, 49
Lin, Justin Yifu, 115
Liu, Daoyu, 79, 80
Liu, Jing, 17
Longxugou slum (play about), 137, 190*n*4
Lyon, Randolph, 126

Mao, Yushi, 116–17, 187*n*25, 187*n*29
Mao Zedong (Mao Tse-tung), 19–20, 41, 97
mass movements (*qunzhong yundong*), 85–86, 88–89, 94, 103, 182*n*5
Mearman, Andrew, 118
Melosi, Martin, 14
Mertha, Andrew, 18
Meteorology Application Institute, 147
migratory birds (storks), 150
mining industry: coal, 28, 31, 142; phosphorus, 75–77; turquoise, 58. *See also* iron and steel industry
Ministry of Education, 45, 80
Ministry of Water Resources and Electric Power, 154
Moser, Adam, 17
Moshui Lake, 140, 149–50, 156
Moulik, T.K., 93
Municipal People's Congress, 146

Nan (South) Lake, 156
National Administrative Litigation Law (1989), 59, 101, 177*n*68
National Bureau of Environmental Protection, 143, 154
National Bureau of Metallurgy, 40, 50, 90
National Chemical Industry Ministry, 50
National Construction Committee, 189*n*51
National Department of Agriculture and Forestry, 97, 98, 99
National Department of Health, 40, 90
National Economic Committee, 124
National Environmental Protection Committee, 154
National Environmental Protection Law (1979): certificate acceleration program, 154; effluent pollution, 146; enactment, 88, 92, 107, 157; revision, 182*n*17

National Environmental Protection Ministry, 154
National Leadership Office of Environmental Protection, 71, 91, 124
National Ministry of Health, 91
natural disasters (floods), 20–21(t), 22, 25(f)
Negru, Ioana, 118
noise pollution, 140–41(f), 149, 152–53, 157, 191*n*23
Number 2 Pharmaceutical Factory, 64–65

People's Liberation Army (PLA), 43, 44, 50, 54, 75, 177*n*58
People's Republic of China (PRC): county vs urban administrative status, 72; economic reforms, 13–14; economists, 105–6; environmental administration reforms, 18; environmental legislation, 15–17, 88, 92, 107, 153–54, 157, 182*n*17; five-year plans, 37, 43, 135, 149, 152, 155, 157–58, 160; foreign relations (Soviet Union), 38–39; grain supply, 26; institutional theory, 37–38; mass movements (*qunzhong yundong*) state campaigns, 85–89, 103, 182*n*5; newspapers, 48; open-door policy, 88; plural institutionalism, 13; population plan, 114; railway system, 20, 38–39, 133(f); Western economic sanctions, 155
Perdue, Peter, 127
Perry, Elizabeth, 14
pharmaceutical industry, 64–65, 176*n*46
Phoenix Hill Examination Hall, 78
Pietz, David Allen, 85
Pigovian taxes, 114
Pingxiang Coal Mining Company, 28, 30
PLA. *See* People's Liberation Army (PLA)
pollution: environmental legislation, 15–17, 88, 92, 107, 153–54, 157, 182*n*17; household waste, 40, 43, 104, 129, 136,

150; media response, 128–29; polluters vs victims, 166; rural vs urban victims, 55–56, 82, 177n62. *See also* air pollution; effluent pollution; industrial waste emissions; noise pollution; soil pollution

pollution rights (right to pollute), 4, 7, 8, 114–15, 120, 126–28, 130, 163, 166, 167

Porter hypothesis, 122

PRC. *See* People's Republic of China (PRC)

proverbs, 37, 134

provincial bureaus, committees, departments, and offices. *See entries starting with* Hubei

Qiaokou district, 142, 146, 148, 153

Qingshan district, 37, 48, 61, 68–70, 179n32, 189n46

Regional Air Force Command, 50

research methodology and records, 5–6

river dolphin (*baiji*), 154–55, 158–59(f), 191n33

rivers. *See* lakes and rivers

"rule-by-law" system (*yifazhiguo*), 16

Rural Development Institute, 109

rural energy. *See* biogas economy

Russo-Japanese War, 31

Ryan, Donald, 127

Sanfei Office (Hubei Provincial Leadership Office of Three Industrial Waste Control), 89, 90–91

Sanfei zhili yundong (Three-Waste Governance Movement), 86, 88–89, 103, 132(t)

sewage interception and treatment projects, 136–37, 149–50, 157–58, 189n46

Shapiro, Judith, 13, 19, 87

Sheng, Chunyi, 30

Sheng, Xuanhuai, 28, 29, 30, 134

Shennongjia forest district, 9, 35

Shi, Tian, 110

Shizhishan (Lion Hill) district, 66

Shizhishan Chemical Factory, 66

Sichuan province, 93–94

Smil, Vaclav, 17, 87, 92–93, 94, 121, 188n38

soil pollution: administrative protocols and warnings, 71; chemical fertilizers and pesticides, 77; cropland contamination, 75–77; letters of appeal, 70–71, 77, 180n50; local reports about, 71, 76–77; mining industry, 71–77, 180n37; public protest, 60, 74; records on, 10; rice paddies contamination, 72–74, 180n39; water supply contamination, 74, 180n44. *See also* agriculture

South-Central University for Nationalities, 67

State Forestry Administration of China, 109

State Planning Commission, 40, 90, 91, 97, 100, 124

Stern, Rachel, 15, 59

Tan, Guofu, 115

Tang, Dr. Ming, 81, 181n57

Tangxun Lake, 46

territorial economics, 109, 111

Thompson, E.P., 114

Three Gorges dam project, 34

Three-Waste Governance Movement (*Sanfei zhili yundong*), 86, 88–89, 103, 132(t)

Tian Ying, 74

Tian'ezhou Yangtze Dolphin National Nature Reserve, 157

Tilt, Bryan, 17

Tongji Medical University, 48, 143

Tort Law (2009), 17

Treaty of Shimonoseki, 31

Index

United States, 15, 126
United States Environmental Protection Agency (US–EPA), 63, 144, 190n10
universities: air-pollution complaints (letters of appeal), 66–68, 179n29; college-entry examinations, 79–80, 180n51; funding, 81; student statistics, 78; water supply, 45–46. *See also* education
urban industrial emissions. *See* industrial waste emissions
urban lakes, 46, 140, 145
US–EPA. *See* United States Environmental Protection Agency (US-EPA)

Vermeer, Eduard, 120–21

water hyacinth, 44
water pollution. *See* effluent pollution
Water Pollution Prevention and Control Law, 146, 147
water supply: contamination, 44–45, 52, 74, 128–29; facilities investment, 45–46, 144, 148; management, 26, 34; wells, 175n28
water-quality monitoring, 48, 106, 140, 142–44, 149–50, 152, 191n17
Wenhua College, 79
wetlands, 22–23, 26, 136
WMA. *See* Wuhan Municipal Archives (WMA)
Worster, Donald, 14
Wuchang district: about, 27, 37, 61, 62; aerial view, 133(f); air pollution, 66–68; colleges and universities, 78–79; intellectual legacy, 77–81, 134; place-name change (Echen/Ezhou/Xiakou), 61; sewage treatment plant project, 136–37; water supply, 137
Wugang Group. *See* Wuhan Iron and Steel Company (Wugang Company/ Wugang Group)

Wuhan: about, 4, 10, 20, 27, 37, 62; aerial view, 133(f); air pollution, 60, 62–70, 178n8, 178n15, 179n18, 179nn21–22; branding as *jianghu*, 84; bridges, 20, 38–39; city plan, 131, 134; climate, 19, 156, 157, 191n39; colleges and universities, 43, 78–79, 81, 174n14; commercial district, 27, 139, 140; cultural legacy, 78, 134–35; effluent pollution, 136–37, 139–40; energy consumption, 142; environment regulations consolidation, 152–58; environmental awareness, 34, 78; environmental monitoring network, 149; environmental protection codes, 139, 151–60; environmental protection score, 154; expansion, 62; factory emissions reports, 62–63, 64, 178n8, 178n15, 179n18; five-tranche categorization system, 152; flood diversion project and urban dikes, 24–25(f); flood incidents, 20–21(t), 22, 137; flood monument and levee, 25(f); industrial emissions surveys, 142, 148; industrial legacy, 27–33; municipal directors conference, 153; noise control devices, 140–41(f); noise decibel levels, 145, 152–53, 191n23; novel about, 156, 191n40; poem about, 20; pollution control devices, 139–41(f), 142; pollution regulation goals, 153–54; pollution-control investment data, 65–66, 179nn21–22; population statistics, 156; proverb about, 134–35; railway system, 20, 38–39, 133(f); sewage treatment plant project, 136–37; streets, 137, 138(f); student statistics, 78, 81; townships, 37; urban lakes, 46; urban zones, 20(f), 43, 147–48; waste discharge fees, 139, 144–45, 152, 155, 157; water management agencies, 34; water supply, 43, 44–46, 143–44, 148;

zones, 20(f), 43, 147–48; zoo, 150
Wuhan Boiler Company, 43
Wuhan Bureau of Environmental Protection (Wugehuan): air-pollution investigations, 65, 178n16; consortium, 146; dust-control zones, 152; monitoring data, 152; pollution-free factories status, 152; research exchange collaboration, 144, 190n10
Wuhan Bureau of Machinery, 95
Wuhan Cement Company, 143, 190n8
Wuhan Changjiang Bridge (Wuhan First Yangtze Bridge), 20, 38, 39
Wuhan Chemical Engineering Company, 147
Wuhan Chemical Material Plant, 146
Wuhan City Zoo, 150
Wuhan Distillery, 153
Wuhan Environmental Research Institute, 147
Wuhan Gedian Chemical Factory, 50
Wuhan Heavy-Duty Machinery Factory, 43, 174n13, 179n18
Wuhan Institute of Chemical Engineering Research, 67, 179n29
Wuhan Institute of Environmental Medicine (Wuhan Tongji Medical University), 48, 143
Wuhan Institute of Hydrobiology, 48, 155
Wuhan Institute of Plant Science, 67
Wuhan Institutes of Iron and Steel, 69
Wuhan Instrument and Meter Factory, 142
Wuhan Iron and Steel Company (Wugang Company/Wugang Group): about, 37, 39; administration, 44, 46, 48; beginnings, 20; dust-control innovation, 143, 190n8; effluent pollution, 46, 142, 179n18; factories, 32(f), 43, 46; mergers, 175n17; warning notice (*xiangai*), 100, 125; worker complaints, 69

Wuhan Military Administrative Region (Military Command of Guangzhou), 43
Wuhan Municipal Archives (WMA), 5–6
Wuhan Municipal Bureau, 154
Wuhan Municipal Chronicle reports: celebrity citizens, 134; dam-dike construction, 151; environmental assessment reports (EARs), 147; environmental protection, 8, 135, 144, 157, 158, 159; infrastructure investment, 140; municipal bureau directors, 153; research institutes projects, 154; river dolphins, 154–55, 157; river-to-road project, 137; sewage treatment plant, 136, 137; urban-rural construction, 135; water facilities statistics, 148; waste-discharge fees, 144–45
Wuhan Municipal Committee of the CCP, 69
Wuhan Municipal Disease Prevention and Monitoring Station (DPMS), 90–92
Wuhan Municipal Environmental Monitoring Network, 149, 191n17
Wuhan Number 2 Chemical Factory, 50, 125
Wuhan Qingshan Sulphuric Acid Factory, 69
Wuhan Textile University, 67
Wuhan University, 48, 79, 80, 154
Wuhan University of Technology, 45, 79

Xi Jinping, 129
Xiaogang county, 76(f), 136
Xiaonan District, 76(f)
Xu, Dixin, 110, 111, 186n10, 187n29
Xunshi River, 149–50

Ya'er (Yaerhu) Lake: covert waste-dumping, 48; dikes, 49; fish

contamination, 49, 90; industrial waste effluent, 46, 48–50; maps, 41(f), 47(f); water-quality monitoring, 48
Yan, Jun, 56–57, 185n1
Yang, Tseming, 16, 17
Yangtze River/Yangzi River. *See* Changjiang (Yangtze/Yangzi) River
Yangtze Water Resources Protection Institute, 147
Yangwu movement, 28, 29, 33, 134
yearbooks and/or chronicles (*nianjian*), 6
Yeh, Emily, 18–19
Yingcheng, 51(f), 52, 53,
Yun River. *See also* Fu (Fuhe) River

Yunmeng, 22, 51(f), 52
Yunmeng Marshland, 22, 23

Zha, Quanxin, 79–80
Zhang, Dr. Peigang, 105
Zhang, Jiayan, 15
Zhang, Zhidong, 27–28, 29, 30, 66, 134
Zhao, Yuhong, 16–17
Zheng, Gongying, 30–31
Zhong, Tianwei, 29–30
Zhonghua University, 79
Zhongnan Materials Research Institute, 95
Zhongnan University of Law and Economics (Hubei University), 66–68, 79, 110, 179n29, 179nn26–27

Printed and bound in Canada by Friesens

Set in Garamond and Times New Roman
by Artegraphica Design Co.

Copy editor: Deborah Kerr

Proofreader: Sophie Pouyanne

Indexer: Margaret de Boer

Cartographer: Eric Leinberger

Cover designer: Will Brown

Cover images:
"Wuhan Urban Map, 1985," *Wuhan Yearbook*
(Wuhan: Wuhan Publishing House, 1986).

Authorized Representative:
Easy Access System Europe – Mustamäe tee 50,
10621 Tallinn, Estonia,
gpsr.requests@easproject.com